Problem-Solving Principles
For Programmers:

Applied Logic,
Psychology,
and Grit

Problem-Solving Principles For Programmers:

Applied Logic, Psychology, and Grit

WILLIAM E. LEWIS
IBM Corporation

HAYDEN BOOK COMPANY, INC.
Rochelle Park, New Jersey

To Barbara, Kim, Eric, and Brian

Who graciously accepted me as a part-time
family member during the writing of this book

Library of Congress Cataloging in Publication Data

Lewis, William E.
   Problem-solving principles for programmers.

   Bibliography: p.
   Includes index.
    1. Electronic digital computers—Programming.
2. Problem solving. I. Title.
QA76.6.L49   519.4   80–23834
ISBN 0-8104-5138-7

Printed in the United States of America

1	2	3	4	5	6	7	8	9	PRINTING
80	81	82	83	84	85	86	87	88	YEAR

Preface

This text has been motivated by a desire to improve problem-solving techniques in computer programming. Its aim is to provide not only a problem-solving background but also alternative solution paths from among which the reader may choose.

Chapter 1 introduces the basic building blocks of problem solving and provides some insights into the psychological influences involved. It sets the stage for the more comprehensive analysis of problem solving developed in subsequent chapters.

Chapter 2 consists of a set of independent "Prescriptions" in problem solving. While these prescriptions apply to general problem-solving situations, their primary purpose is to remedy particular programming problems. Where appropriate, nonprogramming as well as programming problems are presented in order to exercise the reader's mental muscle of analogy.

Chapter 3 consists of a set of advanced "prescriptions" in problem solving to augment the basic prescriptions in Chap. 2.

Chapter 4 presents approaches for attacking more complicated problems for which the prescriptions of Chaps. 2 and 3 may not provide an appropriate panacea. The concept of top-down programming, a relatively new technological development, is the main theme. A programming problem using the top-down approach is illustrated in five different programming languages. A second and more complex problem is also analyzed.

Chapter 5 applies many of the problem-solving techniques discussed in previous chapters for the purpose of eliminating errors, or debugging a program. A set of debugging prescriptions is presented in the fashion of Chaps. 2 and 3.

The entire book consists of three interwoven conceptual threads: general problem solving, program problem solving, and the influence of psychology on the overall problem-solving process. "Silver Bullet Thoughts" are provided to keep the reader on his toes.

The programming examples are given in an "INTERLINGUA" pseudo coding language, but the terminology should be clear even to those without a detailed knowledge of any specific programming language.

There are many people whose talents and influence are reflected in this text, which stems from a course taught at Palm Beach Junior College, Lakeworth, Florida, called "Structured Programming." After teaching this course four consecutive semesters, I realized that although structured programming is a vital tool for improving programming productivity, it falls short as a panacea. Two ingredients appear to be missing—first, the fundamental principles of formal logic, and second, the influence of psychology on the overall problem-solving process. This book attempts to remedy that lack. I thank all students who either directly or indirectly encouraged and influenced me.

I would also like to thank Don Loomis of IBM, Boca Raton, Florida, and Jim Smith of Bell Labs and the University of Chicago for their editing and review. Keith Pandres of Georgia Institute of Technology wrote the PASCAL, COBOL, and FORTRAN programs in Chap. 4. Henry Ledgard of the University of Massachusetts was instrumental in pointing out the uniqueness of such a book as mine and the impact it could have on computer science. His organizational suggestions and encouragement were major factors in the book's completion. Finally, I would like to thank IBM for its aid and assistance.

WILLIAM E. LEWIS

Contents

Problem-Solving Principles For Programmers:

Applied Logic, Psychology, and Grit

1

Framework for Problem Solving

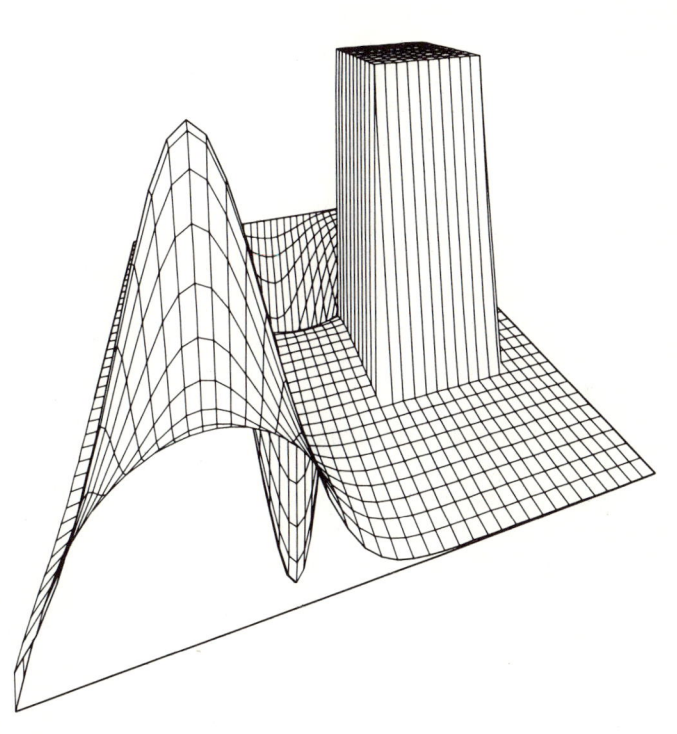

By wisdom a house is built,
and by understanding it is established;
by knowledge the rooms are filled
with all precious and pleasant riches.

Proverbs 24:3:4

1.1 Keys that Open Doors

Imagine that you are inside a large house that contains many rooms and passageways, in one of which is a golden treasure. Since there are so many rooms, many indistinguishable from one another, it is not clear which to enter. To compound the difficulty, entry into some of the rooms requires opening a series of doors in sequence. Some of these doors open easily; others require a great deal of exertion. In addition, some doors are locked whereas others are not. Our intuition tells us that the important rooms are locked and require a "key." Some keys open more than one door, and some doors require more than one key. Worst of all, some rooms are like Pandora's box, for entering them will result in dire consequences.

In our quest for the treasure, random trial and error is the usual approach, but it is an approach that proves time-consuming, for we often become lost in the maze and experience frustration.

The above scenario is typical of the dilemmas we may face when confronted by programming problems. The room containing the treasure is the solution to the problem, and our task is to find this room and unlock the door. A natural tendency in our journey to the treasure, as well as in program problem solving, is to run with the first solution that comes to mind—the door nearest to hand. A far better strategy is to select the most attractive "path" from among many approaches.

Before investigating the various approaches to problem solving, let us first review the steps required to develop a program.

1.2 What Is Programming?

Programming is the task of developing a program—a series of written instructions that make it possible for a computer to solve a problem by accepting data, performing prescribed operations on the data, and supplying the results of these operations. A sequence of well-defined steps is required to develop a good program, and these are shown in Fig. 1.1.

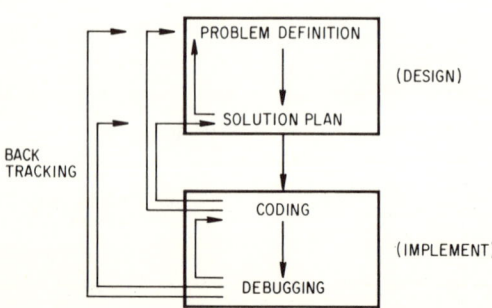

Fig. 1.1 *Program development steps*

2

The first step—problem definition—involves understanding the problem and what is to be done. The problem must be thoroughly thought out before the programmer progresses to the next step in the problem-solving process.

The second step—solution planning—is to develop a plan of attack to solve the problem. Again, this step must be completed in detail before the programmer progresses to the next step.

The third step—coding—is to write the solution developed in the second step in a specific programming language. Computer languages are very similar to the natural languages used by human beings for communicating with one another.

The fourth step—debugging—deals with the verification of the programming logic developed in the planning and coding steps. The purpose of debugging is to insure that the original problem is represented accurately in the programming language and that its solution can be successfully accomplished.

Another step not noted in Fig. 1.1 is documentation. Documentation includes the formal documentation of the logical solution methodology as well as the user procedures required to run a program. It is of fundamental importance in all phases of developing a program since it lets the user know how to run the program as well as how it works. The documented solution methodology may also prove useful to the same programmer or others when attacking new problems.

The first two steps—problem definition and solution planning—deal with the design of the program and specify the problem solution approach. The next two steps—coding and debugging—involve the actual implementation of the solution approach in a language suitable to the computer. They should never be attempted until the first two steps have been carefully thought out. However, it may be necessary to rethink the first two steps because of unforeseen circumstances. For example, additional problem information may be discovered during the actual implementation. If one is careful, however, and exerts sufficient effort in the design phase, backtracking into earlier steps will be minimal. In fact, the thrust of this text is to insist on the fundamental practice of concentrating on proper analysis in the first two steps—problem definition and solution planning. Doing so will assure the following happy results:

1. Less time will be spent on backtracking.
2. When backtracking is necessary, it will not be difficult.
3. There will always be assurance that the problem will be solved.

1.3   The Anatomy of Computer Problems

Let us first recognize that the elements of a general problem and solution approach can be dissected into component parts, just as we can dissect an organism in biology or zoology.

Givens, Operations, and Goals

The first component of a problem is the "givens"—the known facts or assumptions that we accept as present in the world of the problem. The solution to the problem is called the "goal." The vehicle for achieving the goal is a series of one or more "operations"—the actions we perform on the givens to achieve the goal. A visual representation of this simple relationship is shown in Fig. 1.2.

Fig. 1.2   *General problem structure*

The direction of the solution is usually thought of as "forwards" in the sense that we work from the givens until we reach the goal. On certain occasions, however, our approach will involve the reverse; we "work backwards" from the goal to the givens to solve the problem.

A goal in a problem can be classified as (1) incompletely specified, or (2) completely specified. An incompletely specified goal is one whose aim is a general overall strategy rather than a specific result. An example is chess playing, in which the goal is to checkmate the other player in a small number of moves. In this case, the goal becomes more completely specified only as it is approached. A completely specified goal, on the other hand, is clearly and explicitly defined. This implies that all required information is provided in advance. Computer programs are usually characterized as exhibiting completely specified goals, for all information necessary for the desired results is given.

The corresponding elements in a computer to the givens, operations, and goal of a programming problem are input, processing, and output, respectively, as shown in Fig. 1.3.

Fig. 1.3   *Programming problem structure*

Input consists of the facts or data of a problem that have been recorded on a computer-readable medium and transferred into the computer for processing. The medium used to record the data is known as the *input device,* such as a card reader, magnetic tape, optical character reader, magnetic ink character reader, paper tape, terminal, diskette, voice analyzer, or tone analyzer.

Output is the result of the processing of input. The medium used to transmit the output from the computer is called the *output device,* such as a printer, terminal display, magnetic tape, card punch, paper tape, or audio response.

The overall computer programming problem is to manipulate the input in such a manner as to produce the desired output.

1.4   What Are Primitive Thinking Tools?

Primitive thinking tools are the innate thinking processes used in problem solving. The formal recognition of these tools dates back to the Classical Greek period (approximately 600 B.C. to 300 B.C.). These methods for solving problems entail operations that transform the givens of a problem to the goal. We will study the following ones:

1. Working forwards
2. Working backwards
3. Brainstorming
4. Inference
5. Analogy
6. Subgoaling
7. Top-down approach
8. Elimination
9. Contradiction
10. Trial and Error

It should be noted that all these are established techniques for solving problems that can be readily taught. They are not exact, like those of mathematics, but can be looked upon as open-ended, for we have barely scratched their surface. Studying them can be considered a worthy endeavor, for like any other field such as economics, chemistry, or mathematics, they can be considered a single body of important knowledge.

Two Problem-Solving Approaches

There are two major approaches to problem solving. The first, the "behavioristic," concerns itself only with the relationship between a stimulus and response, without analyzing the intervening process. The use of aspirin is a good example of behavioristic problem solving. Aspirin provides effective headache relief even if it is not known exactly how it works.

The second approach, which emphasizes the *process* that occurs between the stimulus and response of a problem, is known as the "information processing approach." It is used in the development of computer programs, and we will examine it in detail.

1.5   Role of Primitive Thinking Tools in Programming

Procedural Logic and Logic

There is some evidence to suggest that programming consists of two distinct parts. On the one hand, it involves rules and procedural elements, such as programming languages, associated grammatical rules, procedures for running a program, etc. This aspect of programming requires the most activity. In addition to these procedural considerations, however, "logical reasoning" must be brought into play to solve any programming problem. Such reasoning is independent of any particular computer or any given programming language, and even of programming itself. It entails the remaining activity.

In the rapidly advancing field of computer science, the procedural aspects of programming have been stressed entirely too much, and the formal study of the fundamental principles of logic has been relatively neglected. For example, when the beginning programming student is introduced to the computer, the major emphasis is placed upon learning a programming language. While this is a very important skill, it often falls short of providing the proper framework for "thinking" in the programming context or developing a program effectively.

Our goal is to apply the primitive thinking tools to programming problems to make up for this neglect. We will make the connection between formal "logic" and program problem solving by using the "teaching by example" approach. Both non-computer and computer examples will be discussed in order to emphasize the generality of each technique.

We will explore the merits of each method and its ability to enhance our journey to a solution. Sometimes our approach to the problem will be through the front door, called "working forwards," and at other times through the back door, called "working backwards." Sometimes we will enter the inner sanctum using a secret passageway.

Our aim is to forge a reasonable (simple, coherent) solution from given requirements, putting aside unnecessary alternative approaches.

1.6   Role of Psychology in Problem Solving

In addition to the fundamentals of problem solving, we should be aware of other factors that contribute to the successful solution of a problem. Psychology, which can be defined as the science dealing with the mind and mental processes, feelings, and desires, intimately influences the problem-solving process and can be instrumental to a problem's solution or failure.

Several psychological processes are integrally related to human problem solving; those that we will explore include the following:

1. Mental blocks
2. Semantics
3. Mentally working forwards
4. Mentally working backwards
5. Psychological closeness to a problem
6. Egoless problem solving
7. Imagination
8. Brainstorming
9. Mental inference
10. Mentally creating additional constraints
11. Mentally relaxing constraints
12. Mental bridging with analogy
13. Mental incubation
14. Ritualistic problem solving and bandwagons
15. Interpersonal communication
16. Mental isolation of a problem key
17. Stress
18. Defensive problem solving
19. Psychology of the game effect
20. Mentally decomposing a problem

As we learn to use our primitive thinking tools, we will investigate how given problems should be approached. We will also see how our mental processes both aid and restrict our ability to solve these problems. This should prove enlightening, for once we understand our limitations, we should be more able to cope with them. We will also see how our journey to the solution is often impaired by our worst enemy, ourselves.

2

Basic Problem-Solving Prescriptions

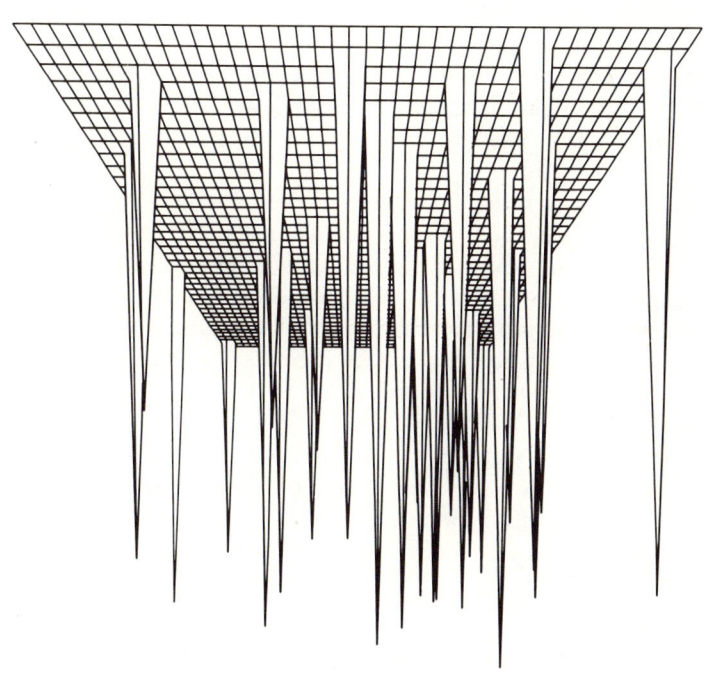

Remedies for bleeding, which fail to check it, are a mockery.

Lewis Carroll (1832–1898)

Introductory Comments

T. H. Huxley, the great nineteenth-century scientist and lecturer, said, "I suppose that medicine and surgery first began by some savage more intelligent than the rest, discovering that a certain herb was good for a certain pain, and that a certain pull somehow or other set a dislocated joint right. I suppose all things had their humble beginnings, and medicine and surgery were in the same condition."[1]

One of the earliest attempts to document methods of cure formally is the Ebers Papyrus, whose "prescriptions" are nothing more than incantations to be recited. In the ancient mind, magic and science were so interwoven that gods and superstition were the ruling forces of medicine. Since the body was considered sacred, no study of anatomy or real analysis of disease was ever undertaken.

Then, in 460 B.C., Hippocrates was born on the island of Cos near the Asclepion. He was destined to become the greatest physician in ancient Greece and came to be known as the "Father of Medicine." Hippocrates set out to learn his subject from actual observation, rejecting all belief in magical practices.

Hippocrates was revered because he was the first man to teach that many diseases clear up best if the physician does not meddle too much and that medicine can advance only when it rejects magic. Disease, said Hippocrates, is a part of the order of nature. Its progress must be watched and can be cured by wholly natural means. Therefore, he insisted that the patient be observed closely, that his changing symptoms be noted day by day, that case histories be recorded, and that cures be "prescribed" accordingly.

One of the most significant writings in the history of medicine is the Hippocratic Oath, which some 25 centuries after it was composed is still subscribed to by physicians all over the world. Its greatest significance lies in the ethical standard, or moral code, that it established and that no reputable physician will transgress. The oath is as follows:

I swear by Apollo Physician, by Asclepius, by Health, by Panacea and by all the gods and goddesses, making them my witnesses, that I will carry out, according to my ability and judgement, this oath and this indenture. To hold my teacher in this art equal to my own parents; to make him partner in my livelihood; when he is in need of money to share mine with him; to consider his family as my own bothers, and to teach them this art, if they want to learn it, without fee or indenture; to impart precept, oral instruction, and all other instruction to my own sons, the sons of my teacher, and to inden-

---

[1] From *On Medical Education in Science and Education,* Appleton, 1894.

tured pupils who have taken the physician's oath, but to nobody else. I will use treatment to help the sick according to my ability and judgment, but never with a view to injury and wrong-doing. Neither will I administer a poison to anybody when asked to do so, nor will I suggest such a course. Similarly I will not give a woman a pessary to cause abortion. But I will keep pure and holy both my life and my art. I will not use the knife, not even, verily, on sufferers from stone, but I will give place to such as are craftsmen therein. Into whatsoever houses I enter, I will enter to help the sick, and I will abstain from all intentional wrong-doing and harm, especially from abusing the bodies of man or woman, bond or free. And whatsoever I shall see or hear in the course of my profession, as well as outside my profession in my intercourse with men, if it be what should not be published abroad, I will never divulge, holding such things to be holy secrets. Now if I carry out this oath, and break it not, may I gain for ever reputation among all men for my life and for my art; but if I transgress it and forswear myself, may the opposite befall me.

HIPPOCRATES (460–377 B.C.)

The present chapter is a collection of independent problem-solving "prescriptions" that will attempt to administer the appropriate medicine to typical programming ills. At first glance, some of the prescriptions may seem trivial or trite and not appear to justify a detailed analysis. However, it is believed that past experience shows that they do indeed provide common-sense panaceas to many of the programming errors consistently repeated every day. If the prescriptions are to be properly applied, the programmer must view himself as a practitioner whose attitude is "practice makes perfect."

PRESCRIPTION 1

Make Sure There Is a Method to Your Madness

One misconception of the nonprogramming community is that programs simply "spring into being"; in reality, nothing could be further from the truth. While programming is not an exact science, a number of clearly identifiable steps are involved in the programming process as discussed in Chap. 1.

Step 1 of the solution process—problem definition—answers questions such as "What is it that I must do?," "Is it clear what my goal is?," and "Do I fully understand what is given?"

To illustrate the difficulties of problem definition, consider the following problem, widely publicized in recent years: How can we solve the energy crisis? At first glance, one might suggest a number of remedial solutions: solar energy, controlled use of energy, or military intervention. However, a fundamental question must be proposed. This question is, "What is the *real* problem?" Is the problem merely that gas prices are skyrocketing, or is it that the earth is physically running out of oil? Is the problem political, or is it the need to develop alternative sources of energy? The questions are endless, but ironically so are the immediate theories that propose solutions.

Obviously, there is a need to define a problem completely before applying problem-solving techniques; to proceed otherwise will yield nothing but wasted efforts. Problem definition involves clearly stating the given information and the goal required.

Often problems are vague and unclear because of "holes," or missing pieces of problem information. In such cases, it may be necessary to create a set of assumptions about these unknowns, and it is important that you identify these assumptions clearly. As additional information is gathered, the assumptions are gradually replaced until any that remain are finally accepted as explicit problem givens. During problem definition, it is necessary to identify the input and output as well as any additional data not specifically given.

Once you have thoroughly understood the problem, you can begin the search process—solution planning—using the trial-and-error method of inspecting the problem to choose your approach. Here you will have to rely on insight. You are really looking for something similar to what you have experienced previously, searching your mind for familiar patterns, asking yourself questions that might reveal a path from the beginning—input—to the end—output. In reality, you are trying to conceptualize an idea. You may know what you want to do, but you need to pinpoint the exact way to do it, excluding those things that have no bearing on the correct solution and concentrating only on the essentials.

The search process shown in Example 2.1 illustrates the trial-and-error methodology. The notation in this example, as well as all future examples, employs an "interlingua" that represents a logic that is independent of any specific programming language (initial asterisks denote comments).

Example 2.1 *Solution Search Process*

```
. . . . . . . . . . . . . . . . . . . . . . . . . . . . . . . . . . . . . . . . . . . . . . . . .
.                                                                  .
.    while solution not found do the following    .
.                                                                  .
.        try something new                                .
.                                                                  .
.    use solution (problem solved)                   .
.                                                                  .
.    end                                                       .
.                                                                  .
. . . . . . . . . . . . . . . . . . . . . . . . . . . . . . . . . . . . . . . . . . . . . . . . .
```

In general, to solve any problem, you follow the sequence of events illustrated in this example. First, you must ask yourself if you recognize the solution or part of the solution or if you recognize a method that will help find the solution. If not, you should try anything, however remote, to see if it helps provide a solution. If it does, then use it. If not, modify it, and try the modified method. Keep modifying and comparing until you have obtained a basic working method.

The important thing in the trial-and-error method is to learn from one's errors and try new variations, different approaches. It is basically a learning situation; you continue to gain new information along the way.

Solution search and planning is thus a more complicated step than problem definition because it involves the generation of a detailed set of specifications and a plan geared to a definite goal.

Unfortunately, solutions are sometimes not properly thought out. As an example, consider the following set of directions for finding a vacation motel:

1. Go south on Highway 6 for about four miles until you pass the drugstore on your right; then turn off at the next intersection.
2. Go along until you pass a maple tree by the road; about a quarter of a mile past the tree look for an asphalt road on the right.
3. Follow the asphalt road until the fourth, or maybe the fifth mailbox, and turn down the driveway.
4. Go down the driveway about 400 yards. The motel is the building with a swimming pool; you can't miss it.

The problem definition is quite clear: Find the vacation motel. However, the instructions for doing so are ambiguous. For instance, it is not mentioned which way to turn at the drugstore. There is no assurance that the maple tree is unique, and it is not clear which mailbox to look for.

When planning a program, we must clearly state the steps involved and precisely describe the solution methodology. Otherwise, we will be faced with the same sort of confusion that we would have in finding the motel. In fact, we are no more likely to reach a solution then we are to reach the motel.

One of the most important processes involved in solution planning is that of properly determining the problem relationships. A relationship can be thought of as being the connection between two or more corresponding things, for example, "father of," "brother of," "heavier than," "equal to," "less than," and so forth.

Computer Examples

The following transportation problem will illustrate problem relationships. With the emphasis on fuel conservation, you are likely to have been monitoring your automobile gas consumption more carefully of late. Assume that your data is that shown in Table 2.1 and that your gas tank has been completely filled each time.

Table 2.1 *Transportation Data*

Date	Odometer Reading	Gallons Added
12/18/76	9201	—
12/27/76	9470	12.5
01/01/77	9550	6.6
01/07/77	9697	9.8
01/11/77	9753	4.5
01/17/77	9844	7.0
01/21/77	9939	7.0
01/30/77	10012	4.0
02/08/77	10121	5.9
02/18/77	10247	7.2
02/22/77	10387	7.5
03/02/77	10515	7.4
03/08/77	10638	7.9
03/15/77	10805	9.0
03/21/77	11043	12.4
03/26/77	11220	9.8

On the basis of this data, you want to determine the following:

1. Miles per gallon for each fill
2. Average miles per gallon
3. Average number of miles between refills
4. Average number of gallons between refills

The problem relationships are as follows:

1. The miles per gallon for each fill is the difference between odometer readings divided by the number of gallons added.
2. The average number of miles per gallon is the total miles traveled divided by the total number of gallons used.
3. The average number of miles between refills is the total number of miles divided by the number of fills.
4. The average number of gallons between refills is the total number of gallons divided by the number of fills.

Example 2.2 shows a program that inputs the data and makes use of the relevant problem relationships to produce the desired output, which is given in Table 2.2.

Example 2.2 *Transportation Problem*

```
PROGRAM: TO COMPUTE MILEAGE STATISTICS

    set TOTAL_MILES      to   0
    set TOTAL_GALLONS    to   0

    print 'Miles Per Gallon'
    input ODOMETER_VALUE, NUM_READINGS
    for CURRENT_READING = 1 to NUM_READINGS do the following

        input NEXT_ODOMETER_VALUE, GALLONS_ADDED

        set MILES_GALLON   to   (NEXT_ODOMETER_VALUE - ODOMETER_VALUE) /
                                GALLONS_ADDED
        print MILES_GALLON

        add (NEXT_ODOMETER_VALUE - ODOMETER_VALUE)   to   TOTAL_MILES
        add GALLONS_ADDED      to   TOTAL_GALLONS
        set ODOMETER_VALUE   to   NEXT_ODOMETER_VALUE

    set AVG_MILES_GALLON   to   TOTAL_MILES / TOTAL_GALLONS
    print 'AVERAGE MILES PER GALLON:',AVG_MILES_GALLON

    set AVG_MILES_REFILL   to   TOTAL_MILES / NUM_READINGS
    PRINT 'AVERAGE MILES PER REFILL:',AVG_MILES_REFILL

    set AVG_GALLONS_REFILL   to   TOTAL_GALLONS / NUM_READINGS
    print 'AVERAGE GALLONS PER REFILL:',AVG_GALLONS_REFILL

end * program *
```

Table 2.2 *Output from Transportation Problem*

Miles per Gallon
21.52
12.12
15.00
12.44
13.00
13.57
18.25
18.47
17.5
18.66
17.43
15.44
18.55
19.19
18.06
Average miles per gallon: 17.03
Average miles between refills: 134.60
Average gallons between refills: 7.90

In solution planning we are also concerned with problem relation-ships from a logical-flow point of view. To illustrate, consider the problem-state diagrams represented in Fig. 2.1. A "state" refers to our particular knowledge of a problem at any given time. This example is limited to only two states to simplify the discussion. Most problems, however, involve quite a few states. State 2 results because of an operation performed on state 1; the direction of change is from state 1 toward state 2. The *effect*, therefore, is the change in state; the *cause* is the operation that produced the change. The *relationship* between state 1 and state 2 is the *operation* that produced the change. Our state of knowledge at state 2 is thus greater than at state 1 because of this operation. In the computer context, an operation is any program input, output, or calculation that changes the state of a program.

Fig. 2.1   *General problem states*

When we are trying to determine the relationships between states, we often find short cuts, as shown in Fig. 2.2. In this example, operations 1 and 2 lead from state 1 to state 3 via intermediate state 2; operation 3, however, leads *directly* from state 1 to state 3.

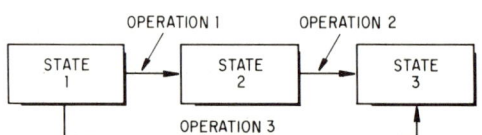

Fig. 2.2   *Problem state short cuts*

Some people are especially good at finding short cuts to a solution. In this case, they might have had a quicker understanding of the equiva-lence of operation 3 to the sequence of operations 1 and 2, but not neces-sarily. Operation 3 may have presented itself only after a diligent pursuit of a solution. For instance, someone intimately familiar with problem relationships may solve a problem only after a long and eloquent series of operations and states. On the other hand, a less experienced person may solve the problem straight off simply *because* he is unfamiliar with the details of intermediate relationships. This is sometimes characterized as "beginner's luck." Nevertheless, experience is essential in problem solv-ing and almost always helps to save time, whether the problem area is a familiar one or entirely new territory.

Determining proper relationships is particularly important in programming because of the need to conserve computer memory. An efficient program has as few instructions, or operations, as possible.

As another illustration of relationships, consider the following simple payroll problem requiring the preparation of payroll stubs for a group of employees. The input for each employee consists of his or her ID, the number of hours worked during a specific week, the hourly pay rate, and the total social security withheld thus far. The output consists of a printed payroll stub. Input and output formats are shown in Table 2.3.

Table 2.3 *Input and Output Formats to Payroll Program*

Input: A sequence of employee cards with the following data:

Data Item	Columns	Format
Employee ID	1–20	ddddddd
Number Hours Worked	22–25	dd.d
Hourly Pay Rate	27–31	dd.dd
Total Security	33–38	ddd.dd

Output: A payroll stub for each employee with the following format:

Col. 5	Col. 15	Col. 25	Col. 35
Line 1→Rate	Hours	Gross	Net
Line 3→dd.dd	dd.d	ddd.dd	ddd.dd
Line 5→Tax	FICA	Other	Employee ID
Line 6→ddd.dd	ddd.dd	ddd.dd	ddddddd

where d = a digit

The first question to be considered is how to compute the gross pay. The usual method is to multiply the number of hours worked by the hourly pay rate. Wrinkles to this computation are introduced by overtime, holidays, and vacations. For example, the number of hours worked over 40 hours may require 1 1/2 times the normal pay rate.

The next step is to compute the income tax. Again, the programmer must clearly state the rules associated with this computation. Fortunately, in this case, the Internal Revenue Service provides tables that explicitly define it.

The next step is to calculate the social security tax. Let us assume that it is 6.5 percent of the gross pay. Note, however, that for this calculation there is also a limit, and once it is reached, the social security tax is no longer withheld.

The final computations are the other deductions, including state income tax, credit union, local income taxes, union dues, bonds, and so forth.

Assume that you have investigated the details of each computation for the payroll problem, which may be described as follows. The gross pay is simply the number of hours worked multiplied by the hourly pay rate. As previously stated, the social security is 6.5 percent of the gross pay. We will assume that the limit on the social security tax to be withheld is $1,200.00. The rest of the deductions are shown in Table 2.4.

Table 2.4 *Deduction Calculations*

(1) *Income Tax*
$100.00 or under: 7 percent of gross amount
Over $100.0: 15 percent of gross amount
(2) *Credit Union*
5 percent of gross amount
(3) *Stock Plan*
10 percent of gross amount

We are now at a point where we can implement the program to solve the payroll problem, as shown in Example 2.3.

You will notice that in the program of Example 2.3 the only considerations other than those previously stated involve controlling the input and processing employee data. These are accomplished by using a loop to read a data record and check for a data termination. The employee stub is then output according to formats specified during problem definition.

Example 2.3 *Program to Process Payroll Stubs*

```
PROGRAM: TO PROCESS PAYROLL

Constant Definitions

   MAX_SECURITY = 1200
   RATE1_INCOME = .07
   RATE2_INCOME = .15
   SOCIAL_RATE  = .065
   CREDIT_RATE  = .05
   STOCK_RATE   = .10
   INCOME_BRACKET 100.0
   NUM_EMPLOYEES = 50

for CURRENT_EMPLOYEE = 1 to NUM_EMPLOYEES do the following

    input EMPLOYEE_ID, HOURS_WORKED, PAY_RATE, TOTAL_SECURITY
    set GROSS_PAY  to   PAY_RATE * HOURS_WORKED

    if GROSS_PAY less or equal to INCOME_BRACKET then
       set INCOME_TAX  to  RATE1_INCOME * GROSS_PAY
    else
       set INCOME_TAX  TO  RATE2_INCOME * GROSS_PAY
```

```
if TOTAL_SECURITY less than MAX_SECURITY then
   set SOCIAL_SECURITY  to   SOCIAL_RATE * GROSS_PAY
   add SOCIAL_SECURITY  to   TOTAL_SECURITY

   if TOTAL_SECURITY greater than MAX_SECURITY then
      subtract (TOTAL_SECURITY - MAX_SECURITY)
              from SOCIAL_SECURITY

      set TOTAL_SECURITY  to   MAX_SECURITY

else

   set SOCIAL_SECURITY  to 0

set CREDIT_UNION  to   CREDIT_RATE * GROSS_PAY
set STOCK_PLAN  to   STOCK_RATE * GROSS_PAY

set OTHER_DEDUC  to   INCOME_TAX + SOCIAL_SECURITY + CREDIT_UNION +
                     STOCK_PLAN

set NET_PAY  to   GROSS_PAY - OTHER_DEDUC

print '   RATE       HOURS       GROSS      NET'
print one blank line

print PAY_RATE, HOURS_WORKED, GROSS_PAY, NET_PAY
print one blank line

print '   TAX       FICA       OTHER     EMPLOYEE ID'
print INCOME_TAX, SOCIAL_SECURITY, OTHER_DEDUC, EMPLOYEE_ID

end * program *
```

Figure 2.3 reviews the basic programming operations and state-change "relationships" of the payroll program. State 2 occurs as a result of inputting the employee name, hours worked, and pay rate. State 3 is reached after the gross pay calculation, and state 4 is reached after the income tax is calculated. As we proceed through the program logic, the current state of the problem changes when we perform new operations and add additional problem information. Our ultimate goal state, however, is to solve the problem completely and arrive at state 8.

In summary, the process of problem solving can be characterized as finding the pathway relationships between the givens and the goal, or input to output. In other words, the methods of problem solving correspond to the ways of finding those permissible paths. In the payroll problem, we planned our solution methodically by first defining the exact input/output format requirements and then proceeding to develop a set of steps or operations to lead us to the goal.

PRESCRIPTION 2

He Who Maketh Haste to Be Rich Shall Not Be Innocent

This prescription basically says that you should start thinking about the problem and lay your plans before attempting any coding, that is, you

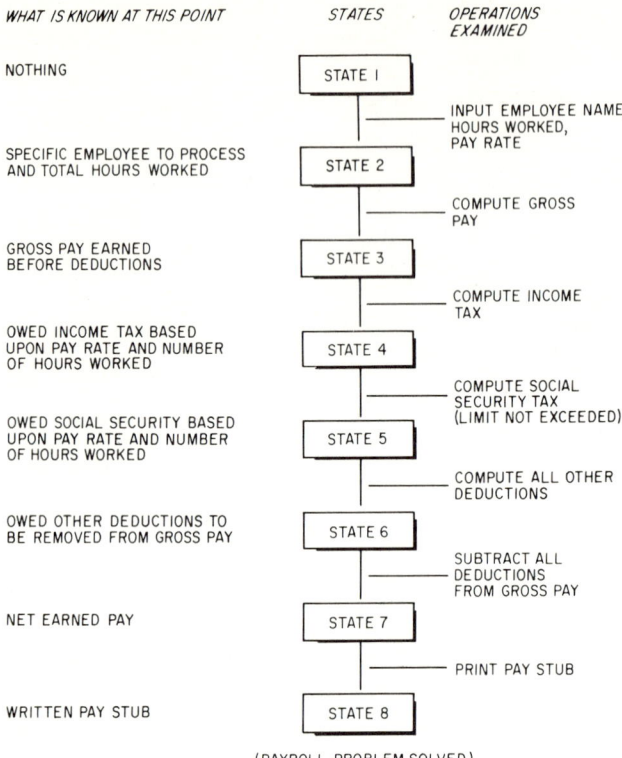

Fig. 2.3   Flow of states in payroll problem

should examine the problem and consider alternative ways to solve it. While this prescription may appear to be obvious, violating it is one of the major culprits in programming, particularly for the beginning programmer. As an illustration of how one can run into difficulty by leaping too quickly, consider the following logical brain teaser.

Imagine that two space ships are moving straight toward each other on a collision course (see Fig. 2.4). One ship is going 8 kilometers per minute; the other, 12 kilometers per minute. Assume that they are 5,000 kilometers apart. What will be the distance between them two minutes before they crash? The typical way to solve this problem is to use algebra, as follows. Let

t = time (2 minutes) before space ships collide

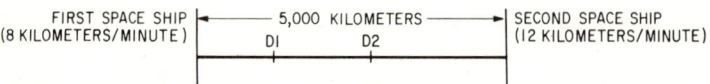

Fig. 2.4   Time scale for space ship problem

d1 = position of first space ship 2 minutes before collision
d2 = position of second space ship 2 minutes before collision

Using the relationship (Rate × Time = Distance), we have:

(1) $8t = d1$
(2) $12t = 5000 - d2$
(3) $(8 \times 2) + (12 \times 2) = d2 - d1$

We now have three equations in three unknowns and can use algebraic elimination to solve for d1 and d2, as follows:

$(3/2)(8t) = (3/2)(d1)$	Multiply both sides of Eq. (1) by 3/2
$12t = (3/2)(d1)$	
$12t = 5000 - d2$	Eq. (2)
$(3/2)d1 + d2 = 5000$	Form new equation from Eq. (2)
$-d1 + d2 = 40$	Collect terms in Eq. (3) for new equation
$(5/2)d1 = 4960$	Subtract terms
$d1 = 1984$	Solve for d1
$d2 = 40 + d1 = 2024$	Express d2 as function of d1

Therefore,

$$d2 - d1 = 2024 - 1984 = \textit{40 kilometers}$$

A much simpler solution is achieved by realizing that the space ships approach each other at 8 and 12 kilometers per minute, respectively. Therefore, the two space ships approach each other at a speed of 20 kilometers per minute, and thus, at 2 minutes before the collision, they must be 40 kilometers apart.

The algebraic approach has three key drawbacks:

1. Overly complicated
2. Unnecessary information
3. Simple solution obscured

First of all, it is more complicated because it requires a number of operations, such as algebraic elimination and substitution. Second, the fact that the two space ships were originally 5,000 kilometers apart is unnecessary information for solving the problem. As a matter of fact, it is this very information that causes most people to approach the problem from an algebraic point of view. Third, the fact that the answer to the problem is given in Eq. (3) of the algebraic formulation was obscured because of our intense concentration on the algebraic operations needed to solve three equations in three unknowns. It thus behooves us to understand the relationships inherent in a problem and to think them out *completely* before attempting a solution.

Computer Example

Let us now consider a computer problem that illustrates the importance of thinking out a problem before coding. Green Thumb brand grass seed costs $3.20 per pound, and one pound will cover 500 sq ft. Write a program that will compute and print the number of pounds of seed, and their total cost, needed to cover areas between 0 and 10,000 sq ft, in steps of 250 sq ft. Your program should read in the cost per pound of the seed and the coverage rate per pound, then print a table such as that shown in Table 2.6.

Table 2.6   *Seed Problem Output*

Area (sq ft)	Amount of Seed (lb)	Cost
0	0.00	0.00
250	.5	1.60
500	—	—
750	—	—
1000	—	—
—	—	—
—	—	—
9500	—	—
9750	—	—
10000	—	—

Since this is a very simple problem, it may seem natural to start coding immediately. The program logic that results is shown in Example 2.4.

Such a program demonstrates the importance of thinking out a problem *before* coding. Three key points are involved here.

Example 2.4   *Agricultural Program (Not Completely Thought Out)*

```
PROGRAM: TO COMPUTE SEED COST (INEFFICIENTLY)

    set AREA_COVERAGE   to   0
    set AMOUNT_SEED     to   0
    print 'AREA IN SQ.FT.   POUNDS OF SEED   COST'
    print '.............   ..............   ....'
    while AREA_COVERAGE less or equal 1000 do the following

        set TOTAL_COST   to   3.20 * AMOUNT_SEED
        print AREA_COVERAGE, AMOUNT_SEED, TOTAL_COST

        add .5 to   AMOUNT_SEED
        add 250 to   AREA_COVERAGE


    end * program *
```

First of all, the problem stated that the cost per pound of the seed and the coverage rate per pound were to be read into the program. In the actual implementation, these two variables were treated as fixed constant data; thus the original problem statement was not completely satisfied. Since the data values are embedded in the program logic, someone not familiar with the program might have difficulty in making changes. For instance, if the cost per pound is to be modified, one would have to know just where in the logic to find this variable.

Second, in our haste to code the program, the area coverage to be tested was given as 1,000 rather than 10,000 sq ft, as the problem required. This clerical error is indicative of the mistakes we can make when we are in a hurry.

Third, when this mistake is corrected, the program logic processes area coverage between 0 and 10,000 sq ft with 250-sq ft increments and explicitly checks for coverage greater than 10,000 sq ft. This check is also imbedded in the logic. While such an approach satisfies the original problem statement, a better program implementation would provide the capability of processing *any* range (to *any* maximum) with a *variable* square-foot increment. This would enable the program to solve any such problem, not just the one stated. A better way of implementing this program would be that illustrated in Example 2.5.

Example 2.5 *Agricultural Program (Completely Thought Out)*

```
PROGRAM: TO COMPUTE SEED COST (EFFICIENTLY)

Constant Definitions

   AREA_INCREMENT = 250
   MAX_AREA = 10000

   set AREA_COVERAGE  to   0
   print 'AREA IN SQ.FT.  POUNDS OF SEED  COST'
   print '............     .............    ....'
   input COST_LB, COVERAGE_RATE

   while AREA_COVERAGE less or equal MAX_AREA do the following

      set AMOUNT_SEED  to   AREA_COVERAGE / COVERAGE_RATE
      set TOTAL_COST  to   AMOUNT_SEED * COST_LB
      print AREA_COVERAGE, AMOUNT_SEED, TOTAL_COST

      add AREA_INCRMENT  to   AREA_COVERAGE

end * program *
```

The area coverage increment is now initialized at 250 and the maximum range at 10,000. In this manner, the values assigned to these variables are located in one place (at the beginning of the program) and can easily be identified and modified. Data values for the cost per pound and coverage rate are initialized by reading and assigning them as variables. The rest of the program deals exclusively with the variables, thus making it more flexible and manageable.

In summary, it is imperative that you think out all problems before starting coding. Everybody who has done any programming knows of the tendency to want to start coding as soon as possible. To do so, however, is like starting out on a trip to unfamiliar territory before considering the best route to take. The only recourse is to develop a discipline that inhibits these "urges" to start programming before a problem is completely thought out.

PRESCRIPTION 3

Keep Hold of Instruction—Do Not Let Go

One of the most common difficulties in computer programming is the failure of the programmer to utilize all information that is necessary for solving the problem.

The first explanation for this failure is the amount of time we sometimes spend worrying about the problem "goal" before thoroughly understanding the problem. Our insistence on attacking the problem immediately, on "getting started," can cause us to ignore some of the given information and the problem relationships. This frame of mind often leads to a more difficult solution or even no solution at all.

Another way to describe this difficulty is in terms of the information-processing models of the brain. The brain, like the computer, can be characterized simply as a combined memory unit and processing unit. The memory unit contains large quantities of information. The processing unit is a short-term memory in which immediate information is processed and manipulated.

The processor must access large quantities of information from the memory unit in a somewhat random fashion. Because of the volume of information involved, the processor unit must "sift" out only that which is relevant. This information is then organized into general categories or "chunks"; thus an organizational process comes into play.

Often a problem may involve a great many givens. The difficulty of keeping track of all this information presents itself when the organizational process becomes laborious. The human processor may not complete the organization of information because it is hard work and also because it is sometimes not very interesting. As a consequence, the organizational process is suspended, and final goals are approached before the problem is properly understood.

Computer Example

Ignoring given information in the problem definition step may be illustrated by the following simple problem that asks us to write a program to keep track of the current balance of a checking account. We are told that the input consists of paychecks, other checks, and special deposits. The output consists of a formatted record of each transaction and the resulting balance, as shown in Table 2.7. The program processing consists of inputting a transaction, determining its type, adjusting the account balance, and formatting and printing the output.

In more detail, the input consists of the following:

1. Transaction type, possibly identified by codes—1000 (deposit), 2000 (paycheck), or a check number.
2. Amount of each check (optional)
3. Amount of each deposit (optional)

One mistake exhibited by many beginning programmers is to check explicitly for each type of transaction. In this case, for example, he would probably explicitly test for checks, deposit checks, and special deposit inputs.

Table 2.7 *Check Balancing Output Requirements*

Item	Amount of Check	Amount of Deposit	Balance
1000		200.15	200.15
6395	49.13		151.02
2000		312.13	463.15
3072	203.01		260.14

A simpler approach that excludes redundant logic is to process transactions according to whether they increase or decrease the balance. The first type of transaction, which increases the balance, includes "deposits" and "paychecks." The second type consists of all other transactions—checks that decrease the balance.

Each transaction is categorized and input as one of these two types. The processing associated with the appropriate category is then executed. A program that illustrates this improved approach, an approach that uses all given information, appears in Example 2.6.

In summary, it is imperative to make advantageous use of all problem information. To do otherwise can make the solution more difficult or even impossible.

Example 2.6   *Check Balancing Program*

```
PROGRAM: TO PROCESS BANK STATEMENT

Constant Definitions

    DEPOSIT_TRANSACTION = 1000
    PAY_TRANSACTION     = 2000
    NUM_TRANSACTIONS    = 100

    set BALANCE  to   0
    print: 'ITEM  AMOUNT OF CHECK  AMOUNT OF DEPOSIT  BALANCE'
    print: '....  ...............  .................  .......'

    for CURRENT_TRANSACTION = 1 to NUM_TRANSACTIONS do the following

        input TRANSACTION_ITEM, AMOUNT

        if TRANSACTION_ITEM equals DEPOSIT_TRANSACTION or
                               PAY_TRANSACTION   then

            add AMOUNT  to  BALANCE
            print TRANSACTION_ITEM, (spaces),AMOUNT, BALANCE

        else

            subtract AMOUNT  from  BALANCE
            print TRANSACTION_ITEM, AMOUNT, (spaces),BALANCE

end * program *
```

PRESCRIPTION 4

Knowledge of Words Is the Gateway to Scholarship

The development of a computer program involves writing instructions that will direct the operations of the computer hardware. The present prescription emphasizes the influence of language on the problem definition step.

Man communicates by means of language. Every language has grammatical rules for generating logical units of thought. These rules must be understood if we are to comprehend fully what is being said. However, even when we fully understand the grammatical structure of a sentence, we may not understand what it is trying to say. In such a situation we must consider the *semantics* involved. Semantics deals with the meaning of words and sentences. Obviously we cannot solve a problem if we don't understand the language in which it is posed. Hence, it is important that we learn to appreciate how semantics can affect problem solving.

Non-Computer Examples

The idiomatic elements of any tongue always present a problem in semantics, especially to foreigners. Since most people are prone to take most things literally, non-English speakers have particular difficulty, for example, when they first encounter an idiom such as "I'm going to give you a piece of my mind." A neurosurgeon not familiar with English would view this idiom with a great deal of alarm! He would have no way of knowing that the expression has nothing to do with any kind of brain surgery, but merely represents the outrage of the individual.

Now consider a cross-word puzzle in which the clue to a four-letter word is "a kind of ruler." Many people might relate this to a straight-edge ruler and have some difficulty in finding the wanted word, "king."

Obviously, if we hope to define a computer program problem properly, we must first clarify what the given information means and the goal we wish to attain. Unless the semantics of the language describing the problem are completely understood, we may waste many hours chasing after the wrong solution.

Computer Example

The following computer programming example shows how semantic difficulties can cause an apparently simple problem to be interpreted in two ways.

The results of a true-false exam given in a class of political science students at the University of Utopia become available for input on cards. Each card contains a student's identification number and his or her answers to ten true-false questions (see Table 2.8).

Table 2.8 Inputs for Exam-Grading Program

Student Identification	Answers (1 = true; 0 = false)
00002	0 0 1 1 1 0 1 0 1 0
00101	1 1 1 1 1 1 1 1 1 1
00205	0 0 0 0 0 1 0 0 0 0
01234	1 0 1 0 1 0 1 0 1 0
10463	1 1 0 0 0 1 0 1 0 0
.	
.	
.	
00045	1 1 0 0 1 1 0 0 0 1
00000	0 0 0 0 0 0 0 0 0 0
The correct answers are:	0 1 0 0 1 0 0 1 1 1

Write a program that will read the data cards, one at a time, and then compute and finally print a two-column table displaying each student's ID and grade. The grade is to be determined as follows: If the score is equal to the best or best minus 1, assign an "A"; if it is best minus 2 or best minus 3, assign a "C"; otherwise, assign an "F."

The semantics problem that we run into here relates to the meaning of the word "best." Does "best" mean the absolute number of correct answers, or 10? Or does "best" mean the best score achieved among all students taking the examination, thus implying grading on a curve?

Example 2.7 shows the program logic to solve the problem when the word "best" is assumed to mean the best score possible. A grade of "A" is thus assigned if a score is 10 or 9; a grade of "C" is assigned if a score is 8 or 7; otherwise, a grade of "F" is assigned.

Example 2.7 *Exam Grading (with Absolute Scale)*

```
PROGRAM: TO COMPUTE GRADES (absolute)

Constant Definitions

    CORRECT_ANSWER(1)  = 0
    CORRECT_ANSWER(2)  = 1
    CORRECT_ANSWER(3)  = 0
    CORRECT_ANSWER(4)  = 0
    CORRECT_ANSWER(5)  = 1
    CORRECT_ANSWER(6)  = 0
    CORRECT_ANSWER(7)  = 0
    CORRECT_ANSWER(8)  = 1
    CORRECT_ANSWER(9)  = 1
    CORRECT_ANSWER(10) = 1
    NUM_RESPONSES = 10
    NUM_STUDENTS  = 50

    print 'STUDENT ID     GRADE'
    for CURRENT_STUDENT = 1 to NUM_STUDENTS do the following

        input STUDENT_ID
        set INDIV_SCORE to 0
        for QUESTION = 1 TO NUM_RESPONSES do the following

            input ANSWER_QUESTION
            if ANSWER_QUESTION = CORRECT_ANSWER(QUESTION) then

                add 1  to   INDIV_SCORE

        if INDIV_SCORE = 9  or  10 then

            print STUDENT_ID, 'A'

        else

            if INDIV_SCORE = 7  or  8 then

                print  STUDENT_ID, 'C'

            else

                print  STUDENT_ID, 'F'


    end   * program *
```

Example 2.8 shows the program logic to solve the problem when the word "best" is assumed to mean the best relative score. A grade of "A" is thus assigned if a score is "best" or "best" minus 1; a grade of "C" is assigned if a score is "best" minus 2 or minus 3; otherwise, a score of "F" is assigned.

Example 2.8 *Exam Grading (with Relative Scale)*

```
PROGRAM: TO COMPUTE GRADES (Relative)

Constant Definitions

    CORRECT_ANSWER(1)  = 0
    CORRECT_ANSWER(2)  = 1
    CORRECT_ANSWER(3)  = 0
    CORRECT_ANSWER(4)  = 0
    CORRECT_ANSWER(5)  = 1
    CORRECT_ANSWER(6)  = 0
    CORRECT_ANSWER(7)  = 0
    CORRECT_ANSWER(8)  = 1
    CORRECT_ANSWER(9)  = 1
    CORRECT_ANSWER(10) = 1
    NUM_RESPONSES = 10
    NUM_STUDENTS  = 50

    set HIGHEST_SCORE  to   0
    print 'STUDENT ID      GRADE'
    for STUDENT = 1 to NUM_STUDENTS do the following

        input STUDENT_ID(STUDENT)
        set INDIV_SCORE(STUDENT)  to   0
        for QUESTION = 1 to NUM_RESPONSES do the following

            input CURRENT_RESPONSE
            if CURRENT_RESPONSE = CORRECT_ANSWER(QUESTION) then
                add 1   to   INDIV_SCORE(STUDENT)

        iF INDIV_SCORE(STUDENT) greater than  HIGHEST_SCORE then

            set HIGHEST_SCORE  to   INDIV_SCORE(STUDENT)

    for STUDENT = 1 to NUM_STUDENTS do the following

        if INDIV_SCORE(STUDENT) = (HIGHEST_SCORE - 1 ) or
                                   HIGHEST_SCORE then

            print STUDENT_ID(STUDENT), 'A'

        else

            if INDIV_SCORE(STUDENT) = (HIGHEST_SCORE - 3) or
                                       (HIGHEST_SCORE - 2) then

                print STUDENT_ID(STUDENT), 'C'

        else

            print STUDENT_ID(STUDENT), 'F'

    end  * program *
```

Depending upon the interpretation of the word "best," one not only has to solve the problem in different ways, one also obtains different results. It is thus imperative that semantic problems be considered when defining a problem, lest one's results fail to coincide with the original problem goal.

PRESCRIPTION 5

Reverse Gears and Work Backwards

Thus far there has been an implied direction in the problem-solving process. This direction is "forwards" in the sense that we work from the givens to reach the goal. Another very useful approach exists that can contribute to a clearer understanding of programming problems. It is called *working backwards*.

In this method, the main focus of attention is the goal to be attained. Starting at the goal, we begin to ask questions like "How can I get to this point?" or "What comes before this point?" We search for preceding conditions that might lead to the goal. From another perspective, consider the concept of cause and effect. In general, a particular cause can be said to produce a particular effect. The reverse concept would be to ask of a given effect, "What was the cause?"

In working backwards, we seek steps that will lead us back to the givens. Figure 2.5 illustrates the concept.

Starting with the goal, we attempt to find some state 1 that may be used with prior information to provide that goal. Then we attempt to find a preceding state 2 that leads to state 1, and so on. We keep on going backwards until we finally arrive at the givens. In a sense, we are searching for the path *from* the givens *to* the goal. But we look for it from the reverse direction.

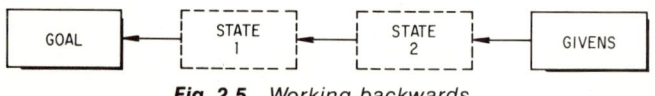

Fig. 2.5 *Working backwards*

Non-Computer Example

As an example of working backwards, consider the problem illustrated by Fig. 2.6. You are provided with three water containers, one a 9-quart, one a 4-quart, and the last a large container into which you are asked to measure exactly 6 quarts from the other two. How can you do this?

Working backwards, we start at the goal and ask the question, "How can we measure out 6 quarts?" Obvious answers are to combine

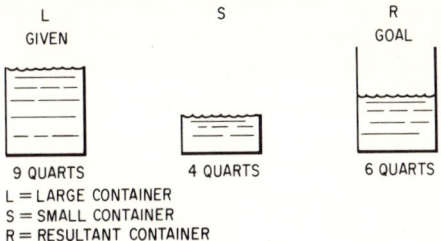

Fig. 2.6 *Water measuring problem*

3 + 3, 4 + 2, 5 + 1, etc. We then ask which of these combinations best relates to the givens? One answer is obviously 4 + 2, since we already have a 4-quart container.

Our problem now becomes, "How can we measure out 2 quarts?" We will soon find that this is a difficult task and we must seek another solution. If we can somehow measure out 3 quarts, we need only repeat the process to obtain the desired 6 quarts. The goal and previous states are illustrated in Fig. 2.7.

Fig. 2.7 *States in working backwards*

In order that we might illustrate the process more clearly, consider the nomenclature in Table 2.9.

Table 2.9 *Terminology of Logical Operations*

S represents the small container
L represents the large container
R represents the goal container

Let S→0 = Empty the small container
     L→0 = Empty the large container
     S→4 = Fill the small container to top
     L→9 = Fill the large container to top
     L into S = Try to pour the large container into the small
     S into L = Try to pour the small container into the large
     S into R = Try to pour the small container into the goal container
     L into R = Try to pour the large container into the goal container

---

One way to measure out 3 quarts is shown in Table 2.10. The goal container will contain 3 quarts at the completion of these operations. When the very same process is repeated, the goal container will contain 6 quarts, as we wanted.

Table 2.10   *Logical Operations to Solve Water Measuring Problem*

Current Contents of Containers		Logical Operations
Small	*Large*	
0	0	L→0, S→0
4	0	S→4
0	4	S into L
4	4	S→4
0	8	S into L
4	8	S→4
3	9	S into L
0	9	S into R

Notice that the method we used for keeping track of the current contents of the containers in Table 2.10 is at the same time a debugging method for verifying our process. The same sort of technique is used for debugging a computer program; in this case, the current values of the variables are kept track of as we desk-check the program. More will be said about this matter in Chap. 5 when we discuss working backwards in the context of program debugging.

Numerous other examples might be cited to illustrate working backwards, but the important thing to remember is that it represents a direction *away from* the goal and *towards* the givens. Often the number of paths leading from the givens is enticingly large, but the paths leading directly to the goal are limited or blocked. In such cases, the search for the correct path may be more easily pursued by working backwards.

Computer Example

As an illustration of working backwards in computer programming, consider the following problem. For the purchase of a new car, you need to write a computer program that will compute your total interest costs and the length of time you will have to pay the different monthly payments. You also want to calculate the amount of each payment that goes toward interest, the amount that goes toward paying off the principal, and the amount of the final payment. The original loan amount is $5,000, and the interest is 1 percent. The desired outputs are shown in Table 2.11, in which "Month" = current month, "Balance" = balance of loan by month, "Interest" = interest paid by month, "Principal" = principal paid by month, "Payment" = monthly payment, "Total interest" = cumulative interest paid to date, and "Total Paid" = cumulative amount paid to date. The monthly payment is $250.00, and the last payment is $106.74.

Table 2.11 *Interest Problem Output*

Month	Balance	Interest	Principal	Total Interest	Total Paid
1	4800.00	50.00	200.00	50.00	250.00
2	4598.00	48.00	202.00	98.00	500.00
3	4393.98	45.98	204.02	143.98	750.00
4	4187.92	43.94	206.06	187.92	1000.00
5	3979.80	41.88	208.12	229.80	1250.00
6	3769.60	39.80	210.20	269.60	1500.00
7	3557.29	37.70	212.30	307.29	1750.00
8	3342.87	35.57	214.43	342.87	2000.00
9	3126.30	33.43	216.57	376.30	2250.00
10	2907.56	31.26	218.74	407.56	2500.00
11	2686.63	29.08	220.92	436.63	2750.00
12	2463.50	26.87	223.13	463.50	3000.00
13	2238.14	24.64	225.37	488.13	3250.00
14	2010.52	22.38	227.62	510.52	3500.00
15	1780.62	20.11	229.90	530.62	3750.00
16	1548.43	17.81	232.19	548.43	4000.00
17	1313.91	15.48	234.52	563.91	4250.00
18	1077.05	13.14	236.86	577.05	4500.00
19	837.82	10.77	239.23	587.82	4750.00
20	596.20	8.38	241.62	596.20	5000.00
21	352.16	5.96	244.04	602.16	5250.00
22	105.68	3.52	246.48	605.68	5500.00
23	0.00	1.06	105.68	606.70	5606.74

There are many such computer programming problems in which the output or goal is initially well established. We now wish to work backwards in the problem-definition and solution-planning steps of the problem.

The primary task in the problem-definition step is to define the input and output clearly. Our immediate goal is therefore to determine the inputs required to achieve the specified output, but we should first understand the calculations required. From simple economics, it is clear that these are as follows:

(1)   Interest (i) = Rate × Balance (i − 1)
(2)   Principal (i) = Payment − Interest (i)
(3)   Balance (i) = Balance (i − 1) − Principal (i)
(4) Total interest (i) = Interest (i) + Total interest (i − 1)
(5)   Total paid (i) = Total paid (i − 1) + Principal (i) + Interest (i)

where:

> Rate = interest rate
> i = current month (1, 2, 3 . . .)

Equation (1) states that the interest for the current month will be calculated as the fixed monthly interest rate times the loan balance from the previous month. For the first month, the balance used to calculate the interest will be the original loan amount, or Balance(0). Therefore, we know that Balance(0) and the fixed interest rate are two inputs required by the program.

Equation (2) states that the principal for the current month will be equal to the monthly payment minus the interest taken out for the month. Therefore, the monthly installment payment is also be an input required by the program.

By inspection, it is obvious that equations (3) through (5) can be calculated from available data and require no additional input.

In summary, the inputs required by this program are as follows:

> Rate = Monthly interest rate
> Balance (0) = Original loan amount
> Payment = Monthly payment

The only other consideration is the payment for the last month. If balance (i) in Eq. (3) for month i is less than zero, then this is the last month and we have the following relationships:

> Balance (i) = 0
> Principal (i) = Balance (i − 1)
> Payment = Interest (i) + Principal (i)

We have thus started from the output goal to define the input of the program and have also defined the intermediate computations required. By working backwards we have developed a clear understanding of the problem. The program that results is shown in Example 2.9.

Review of Basic Idea

Working backwards is characterized by focusing on the goal rather than the givens as the starting point of the problem-solving process. We attempt to guess the conditions that will lead to the goal desired.

Generally speaking, this approach is recommended under two circumstances. The first presents itself when a problem offers a goal that is

Example 2.9 *Interest Program*

```
PROGRAM: TO COMPUTE LOAN INTEREST

Constant Definitions

  NUM_MONTHS = 24

  set TOTAL_INTEREST(0)  to  0
  set TOTAL_PAID(0)  to  0
  input RATE, LOAN, PAYMENT
  set BALANCE(0) to LOAN
  print 'MONTHLY PAYMENT:',PAYMENT
  print one blank line
  print '                              TOTAL    TOTAL'
  print 'MONTH BALANCE INTEREST PRINCIPAL INTEREST PAID'
  print '..... ....... ........ ......... ........ .....'

  for MONTH = 1 to NUM_MONTHS do the following

      set INTEREST(MONTH)  to   RATE * BALANCE(MONTH-1)
      set PRINCIPAL(MONTH) to   PAYMENT - INTEREST(MONTH)
      set BALANCE(MONTH)   to   BALANCE(MONTH-1) - PRINCIPAL(MONTH)

      if BALANCE(MONTH) less than 0 then

        set  BALANCE(MONTH)    to  0
        set  PRINCIPAL(MONTH)  to  BALANCE(MONTH-1)
        set  LAST_PAYMENT  to  INTEREST(MONTH) + PRINCIPAL(MONTH)
        set  PAYMENT  to  0

      set TOTAL_INTEREST(MONTH)  to INTEREST(MONTH) +
                                    TOTAL_INTEREST(MONTH-1)

      set  TOTAL_PAID(MONTH)  to  TOTAL_PAID(MONTH-1) +
                                    PRINCIPAL(MONTH) + INTEREST(MONTH)

      print MONTH, BALANCE(MONTH), INTEREST(MONTH), PRINCIPAL(MONTH),
            TOTAL_INTEREST(MONTH), TOTAL_PAID(MONTH)

  print 'LAST PAYMENT:', LAST_PAYMENT

end  * program *
```

sufficiently well defined for us to concentrate on it and work backwards to the givens. The second depends upon the nature of the operations involved in the problem. If they are one-to-one, working backwards can be quite effective. One-to-one operations are those in which a specified output is generated by a uniquely defined set of inputs, as was the case in the preceding interest problem.

PRESCRIPTION 6

Step Back and View the Forest

Often we become so involved with a programming problem that we cannot see the overall picture no matter how hard we try. This difficulty is characteristic of problems involving many variables.

Fig. 2.8 *Problem in perception*

Perhaps the best way to illustrate this difficulty is to tackle a problem in perception. Can you tell what is illustrated in Fig. 2.8? If you are like most people, you immediately see a vase. Obvious as the vase is, however, it hides another picture. If you do not see the other picture, try looking at Fig. 2.9, which shows the reverse—a dark vase on a white background. Do you see the other picture now? If not, don't be upset, for reversing the background rarely helps in such situations, and most people still will not see the other picture.

Fig. 2.9 *Problem in perception (cont'd)*

Perhaps the difficulty lies in the fact that the vase is so dominant an image that it causes us to ignore any other possibility. The same can also be true of problem solving, where certain aspects of a problem can obscure more subtle problem relationships. Figures 2.8 and 2.9 are examples of a well-known visual effect described by the psychologist, Edgar Rubin. The phenomenon, called "figure-ground reversal," occurs when perception may fluctuate between two possibilities and demonstrates how a problem may be viewed from more than one perspective. If you still cannot see the other picture, we will now point it out. The other picture is that of a pair of faces looking at each other!

Another source of difficulty in perception (or problem solving) is being too close to a picture (or problem). By concentrating too hard on specific areas and being confronted with many isolated images, you lose sight of the whole picture (or problem). The source of this difficulty is called "tunnel vision," in which your concentration on *details* prevents you from seeing the forest for the trees.

The present prescription says that you should not allow yourself to be blinded by glaring facets of a problem and should step back from it when experiencing difficulties. To do otherwise may leave you confused and frustrated. It may be that you are trying too hard to solve the problem or are looking for something that is not there. On the other hand, the problem may be so complex that you must take measures to gain a broader perspective.

PRESCRIPTION 7

He Who Digs a Pit Will Fall into It

One of the most important aspects of program problem definition and solution planning is determining the constraints of a problem. In fact, it is fundamental. Nevertheless, it must also be understood that placing too many constraints can prove to be as fatal as not placing enough. In either case, we may make a solution difficult if not impossible to obtain.

In addition to inputs, operations, and the goal, constraints are basic. They are those limitations placed on a problem that inhibit us from proceeding from the givens to the goal (see Fig. 2.10). The constraints may be either clearly defined or merely implied by the givens.

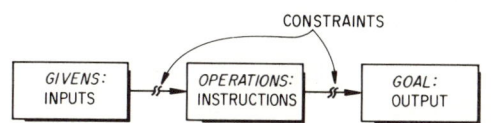

Fig. 2.10 *Constraints to a programming problem*

Before attempting to find the method or operations necessary for solving a problem, you must determine all the limitations to be considered. This process may not be easy; in fact, it may be the hardest job of all, but under no circumstances may it be avoided.

You must first ask the question, "How can I determine the constraints of the problem at hand?" The answer is to study the givens until they are completely clear. You may also need to ask the questions, "Is it clear what my goal is?" and "Do I fully understand what is given?"

Another possible question is whether there may be constraints that are inferred rather than stated. We shall discuss the method of determining such constraints a little later.

As an example of natural constraints, consider the following scenario, illustrated pictorially in Fig. 2.11.

Fig. 2.11 *String problem*

Problem Scenario

Givens:	Two strings hanging from a ceiling and a pair of scissors
Problem:	Tie the two strings together
Natural constraint:	The strings are too far apart to hold one string and walk over and grab the other

The problem at first glance seems to be unsolvable, for we are physically constrained by the length of the two pieces of string. The scissors do not seem to be an adequate instrument for connecting the strings. As a matter of fact, scissors are associated with "cutting," and it is not apparent how cutting strings can aid in the solution.

After further contemplation, however, it becomes obvious that the scissors *must* have something to do with the solution. With still more thought, we realize that if we tie the scissors to one of the strings and swing it so that it comes to us while we are holding the other, the original natural constraint is overcome. Note here that previous experiences with scissors have stereotyped their function and placed an unnecessary constraint on the solution of the problem.

Another example of unnecessary constraints is provided by the following river crossing problem (see Fig. 2.12).

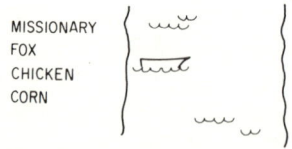

MISSIONARY
FOX
CHICKEN
CORN

Fig. 2.12 *River crossing problem*

A missionary, a fox, a chicken, and a bag of corn are to be transported across a river by a single boat. The boat will hold a maximum of only two items in any one crossing, and one must be the missionary.

The constraints in this problem are as follows. We cannot leave the chicken and the fox alone together since the fox would eat the chicken.

Also, we cannot leave the chicken alone with the corn since the chicken would eat the corn.

In order to resolve this difficulty, we must realize that the basic constraint is the location of the chicken. It cannot be left with either the fox or the corn. Hence, the chicken must be moved constantly so as not to be left alone with either of the other two. As a result, extra boat trips will be required to deal effectively with the constraint imposed by the chicken. The solution to the problem is illustrated by Fig. 2.13.

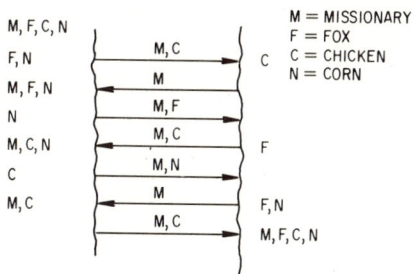

Fig. 2.13 *Solution to river crossing problem*

First, the missionary transports the chicken to the other side. Then he can arbitrarily pick up either the fox or the corn and transport it to the other side, returning with the chicken. The chicken is left on the original side, while he transports the other object to its destination. After doing this, he returns and transports the chicken once again.

Notice that the chicken must be taken across the river three times to prevent it from being left alone with either the fox or the corn. This is the price we must pay to handle the constraint it imposes. If we resist making these extra trips, which seem unnecessary, in order to solve the problem, our reluctance will increase the time the problem requires.

If you attempt, either consciously or unconsciously, to solve the transportation problem by using the least number of trips, you add a constraint that is not part of the problem. Incidentally, this additional constraint greatly increases the degree of difficulty of the problem and might even make it impossible to solve. Unnecessary constraints like this one tend to be inherent in many problems, including programming problems.

Such constraints in the context of programming lead to "tricky programming." Tricky programming results from the attempt of programmers to develop a clever solution that will conserve memory storage or impress their colleagues. The difficulty with such programs is that they tend to confuse the programmers who must maintain them, resulting in wasted time and even the introduction of errors.

A second type of unnecessary constraint can be introduced by trying to solve a problem in the least amount of time—a common enough tendency. As a matter of fact, our society promotes such behavior, which results in problems being insufficiently analyzed and often impedes the understanding of the genuine contraints.

Our impatience is often due to a lack of interest as well as a lack of time. Impatience will definitely impede problem definition. The only obvious remedy is to become more interested in the solution process. When accelerated programming schedules contribute to the lack of time, the usual remedy is to work overtime or to increase the manpower, but the jury is still out regarding the advantages of these approaches.

In addition to the constraints discussed, the use of the computer itself may produce still others. These manifest themselves in two forms: software and hardware.

Software consists of the executable computer instructions that direct the operations of the computer hardware. The degree of flexibility of the software can be constrained by hardware capabilities. For example, the amount of memory storage available is one of the most crucial constraints to any computer program, unless, of course, the storage is unlimited. This constraint may cause you to consider the use of secondary storage (such as disks) as an intermediate facility or even to limit the scope of the original problem.

Other hardware constraints include the speed of the central processing unit (CPU) and the input/output devices. CPU speed becomes a constraint if it is incapable of processing programs in a sufficiently timely fashion. For example, CPU speed may hamper the assembling and compiling of programs. Input/output timing constraints manifest themselves when I/O devices are not fast enough to input or output data at the demanded rate.

A final hardware constraint is limited input/output capacity, which prevents the devices from storing the volume of data dictated by the problem. The original programming goal may have to be redefined to accommodate this hardware constraint.

Constraints may be inherent in a programming problem or dictated by economics. If flexibility exists in the economic area, hardware constraints should be minimized whenever possible.

Software constraints include the level of the programming language as well as the specific language being used. The characteristics or capabilities of any one language can definitely affect the overall computer solution. The fact that you must constantly be aware of computer hardware when writing a program in machine language can distract you from the problem-solving process and become a constraining factor in itself.

On the other hand, the language you are using may not have all the computational features required by your program. This constraint is usually solved by creating the needed features within your program.

Computer Example

The following scenario demonstrates the programming difficulties encountered when unnecessary constraints are employed. Imagine that you are a newly hired programmer for the Acme Company, which has just installed a powerful mini-computer system. During your interview, management stressed their interest in developing a wide variety of personnel reports with the new computer.

It is soon announced that all employees with 25 or more years of service will be recognized at the company's annual awards luncheon. Your first job is to prepare a list of employees fulfilling that condition. Your program is to read in personnel data records, select all those indicating the required years of service, and print a report. The output format is shown in Table 2.12. The report is to terminate when the end of the data is detected.

Table 2.12 *Company Service Report Formats*

Print Positions	Field Description
1–20	Employee ID
22–23	Years of service
25–34	Department number

The personnel data is placed on records with the format shown in Table 2.13 (d indicates a digit).

In a short period of time you develop the program logic shown in Example 2.10.

Management is pleased with your work and realizes the power and efficiency of the new computer. Shortly, it presents you with requests for different types of reports. The first entails a breakdown of employees by department. Just as you are about to complete it, management requests another to provide a breakdown of employees by age groupings for use in a pension plan. Again, before completing this program, you receive additional requests, for example, a report to list all male employees less than 30 years of age whose income is at least $20,000.

By this time, you realize that your approach to writing reports is getting out of hand even though each program is functionally simplistic. It is bad enough having to write a separate program for each report, for there is a common logic in all, but you must also maintain the whole lot and rerun them periodically.

A friend of yours with quite a lot of experience in program problem solving suggests a new approach that appears to be much more flexible, namely, to prepare a program with the capability of producing *any* report based on any or all personnel factors.

Table 2.13 *Personnel Input Record Format*

Card Columns	Field Description
1–20 ddd . . .d	Employee ID
22–23 dd	Age
25 d (1 = Male 2 = Female)	Sex
27–36 ddd . . .d	Department number
38–43 dddddd	Date of employment (Day/month/year)
45–53 ddddd .dd	Yearly salary
55–56 dd	Position grade

Table 2.14 *First Header Card Input Format*

Card Columns	Field Description
1–20 ddd...d	Employee ID
21–25 dd–dd	Age range
26 d (1 = Male 2 = Female)	Sex
27–36 ddd...d	Department number
37–49 dddddd–dddddd	Date-of-employment ranges (Day/month/year)
50–68 dddddd.dd–dddddd.dd	Yearly salary range
69–73 dd–dd	Position grade range

Example 2.10   Personnel Report

```
PROGRAM: TO COMPUTE PERSONNEL REPORT (narrow approach)

Constant Definitions

    DAY_CURRENT      =   15
    MONTH_CURRENT    =   12
    YEAR_CURRENT     =   79
    RETIRE_YEARS     =   25
    NUM_EMPLOYEES    =  100

    print 'YEARS SERVICE REPORT'
    print '....................'
    print one blank line
    for Current_EMPLOYEE = 1 to NUM_EMPLOYEES do the following

        input EMPLOYEE_ID, AGE, SEX, DEPARTMENT, DAY_EMPLOYMENT,
              MONTH_EMPLOYMENT, YEAR_EMPLOYMENT, SALARY, POSITION

        set YEARS_SERVICE  to  YEAR_CURRENT - YEAR_EMPLOYMENT

        if MONTH_EMPLOYMENT greater than MONTH_CURRENT then

            subtract 1 from YEARS_SERVICE

        else

            if MONTH_EMPLOYMENT = MONTH_CURRENT and
               DAY_EMPLOYMENT greater than DAY_CURRENT then

                subtract 1 from YEARS_SERVICE

        if YEARS_SERVICE greater or equal RETIRE_YEARS then

            print EMPLOYEE_ID, YEARS_SERVICE, DEPARTMENT

end   * program *
```

The design of such a program requires that two header records be positioned ahead of the existing input to define the type of report desired. The first record is shown in Table 2.14 (d indicates a digit).

The format is similar to that in Table 2.13 with two exceptions. First, ranges for age, date of employment, yearly salary, and position grade are allowed. For example, employees of age 25 through 30 can be selected and reported. Second, any field in this header record can be left blank by instructing the program to ignore it. For example, if the sex field is blank, the report will treat males and females equally.

The second record card contains an 80-column alphanumeric field for the report title.

Use of the two header records allows reports to be prepared with only *one* program. Personnel data records can be processed in any way merely by changing the header records. The logic for this master program is shown in Example 2.11.

The preceding scenario typifies the difficulties encountered when unnecessary constraints invade program problem solving. If you recall,

43

Example 2.11   *Personnel Report (Master Version)*

```
PROGRAM: to COMPUTE PERSONNEL REPORT (general approach)
Constant Definitions

  NUM_EMPLOYEES  = 100
  NUM_CHARACTERS = 80
  **
  ** main program
  **
  **
  input BASE_EMPLOYEE_ID,LOW_AGE,HIGH_AGE,BASE_SEX,
        BASE_DEPARTMENT,LOW_DAY_EMPLOYMENT,LOW_MONTH_EMPLOYMENT,

        LOW_YEAR_EMPLOYMENT,HIGH_DAY_EMPLOYMENT,
        HIGH_MONTH_EMPLOYMENT,HIGH_YEAR_EMPLOYMENT,LOW_SALARY,
        HIGH_SALARY,LOW_POSITION,HIGH_POSITION

  for CHAR_POSITION equals 1 to NUM_CHARACTERS do the following

      input HEADER(CHAR_POSITION)

  for CHAR_POSITION equals 1 to NUM_CHARACTERS do the following

      print HEADER(CHAR_POSITION)

  for CURRENT_EMPLOYEE equals 1 to NUM_EMPLOYEES do the following

      input EMPLOYEE_ID, AGE, SEX, DEPARTMENT, DAY_EMPLOYMENT,
            MONTH_EMPLOYMENT, YEAR_EMPLOYMENT, SALARY, POSITION

      set  MATCHED_DATA  to  true
      call CHECK_EMPLOYEE_ID
      call CHECK_AGE

      call CHECK_SEX
      call CHECK_DEPARTMENT

      call CHECK_DATE_EMPLOYMENT
      call CHECK_SALARY
      call CHECK_POSITION

      if MATCHED_DATA true then
         print EMPLOYEE_ID,' ',AGE,' ',SEX,' ',DEPARTMENT,
               ' ',DAY_EMPLOYMENT,MONTH_EMPLOYMENT,

               YEAR_EMPLOYMENT,' ',SALARY,' ',POSITION

  subroutine CHECK_EMPLOYEE_ID
     if (BASE_EMPLOYEE_ID not equal 0) and
        (EMPLOYEE_ID not equal BASE_EMPLOYMENT_ID) then

        set MATCHED_DATA  to  false
     return
  end  * subroutine *

  subroutine CHECK_AGE
     if ((AGE less than LOW_AGE) or (AGE greater than HIGH_AGE)) and
        (LOW_AGE not equal 0) then
        set MATCHED_DATA  to  false
     return
  end  * subroutine *
```

```
subroutine CHECK_SEX
    if (SEX not equal BASE_SEX) and (BASE_SEX not equal 0) then

        set MATCHED_DATA  to   false
    return
end  * subroutine *

subroutine CHECK_DEPARTMENT
    if (DEPARTMENT not equal BASE_DEPARTMENT) and
       (BASE_DEPARTMENT not equal 0) then

        set MATCHED_DATA  to   false
    return
end  * subroutine *

subroutine CHECK_DATE_EMPLOYMENT
    set DAYS_LOW to LOW_YEAR_EMPLOYMENT       * 365 +
                    LOW_MONTH_EMPLOYMENT      * 30  +
                    LOW_DAY_EMPLOYMENT
    set DAYS_HIGH to HIGH_YEAR_EMPLOYMENT     * 365 +
                     HIGH_MONTH_EMPLOYMENT    * 30  +
                     HIGH_DAY_EMPLOYMENT
    set DAYS to YEAR_EMPLOYMENT               * 365 +
                MONTH_EMPLOYMENT              * 30  +
                DAY_EMPLOYMENT

    if (DAYS less than DAYS_LOW or
        DAYS greater than DAYS_HIGH) and
        LOW_DAY_EMPLOYMENT not 0 then

        set MATCHED_DATA to fALSE

    return
end  * subroutine *

subroutine CHECK_SALARY
if ((SALARY less than LOW_SALARY) or
    (SALARY greater than HIGH_SALARY)) and
    (LOW_SALARY not equal 0) then

    set MATCHED_DATA  to   false
return
end  * subroutine *

subroutine CHECK_POSITION
if ((POSITION less than LOW_POSITION) and
    (POSITION greater than HIGH_POSITION)) and
    (LOW_POSITION not equal 0) then

set MATCHED_DATA  to   false
return
end  * subroutine *

end  * program *
```

management stated at the outset that they intended to produce a variety of reports. Instead of developing a generalized program to handle all reports, you began by writing a separate program for each report. In other words, you did not think the situation out carefully, and your initial approach constrained programming productivity by restricting the ease of creating new reports.

PRESCRIPTION 8

Happy Is the Man Who Seeks Understanding

One of the most important primitive thinking tools is *inference,* the drawing of conclusions from evidence, or to put it in a slightly different way, looking for other meanings or interpretations. Inference is one of the first problem-solving methods to resort to because it is basically a method of generating additional information from that already given.

Problems often evolve from a vague to an explicit formulation. The source of the vagueness is the *implicit* information, that is, the information not actually stated.

To illustrate a subtle example of inference, consider the following Sherlock Holmes scenario. Holmes handed a hat to Watson and asked him what he could infer about its owner. After examining it carefully, Watson said, "Something about the character of the man is clearly indicated, but it is one of those facets of his spirit that cannot be expressed in words." Hiding his impatience, Holmes asked Watson what evidence led him to this conclusion. Watson replied, "It is an ineffable quality about the hat—something I couldn't possibly describe."

Two criteria determine whether inference should be employed. The first one is the existence of similar situations involving the same type of information. We should therefore look for analogies in our previous experiences. The second is when the additional information generated by inference relates directly to the information provided in the givens and the goal of the problem.

Non-Computer Example

The following example will illustrate how effectively inference can be used to solve a complex problem—the Three Hats Problem.

Three men are condemned to die, one of whom is blind. The King decides that he will offer them an opportunity to be set free. He arranges them in a circle facing one another and produces five hats, three white and two black. He places one hat on each of their heads and destroys the remaining two hats. He then tells them, "The first one of you who can tell me the color of his hat will be set free." A period of time passes. Finally, the blind man tells the King the color of his hat and how he knows what it is. He is set free. What was the color of his hat and how did he know?

The process of inference is vital in solving this problem. The only clue we are given is that a period of time passes before the blind man speaks to the King. What does this pause infer? First of all, it infers that neither man who could see was able to be sure of the color of his own hat.

To analyze why neither of them could be sure, let us construct a chart that illustrates the possible combinations of hats on all three men (Table 2.15).

Table 2.15 *Three Hats Problem—List of Possibilities*

Case	Blind Man	Second Man	Third Man
1	W	W	W
2	W	W	B
3	W	B	W
4	W	B	B
5	B	W	W
6	B	W	B
7	B	B	W

Notice cases 6 and 7. In case 6, the second man would see two black hats, while in case 7 the third man would see two black hats. We can therefore infer the following. If either man saw two black hats, he would then know his own was white, since there are only two black hats. Since both remained silent, we now are able to eliminate cases 6 and 7. Notice that of the remaining five cases, only case 5 puts a black hat on the blind man's head. If case 5 can be proven invalid, we would know that the blindman's hat is white.

Consider our conclusion that neither sighted man saw two black hats. We can state this another way. Each of them must have seen one or two white hats, but *at least one*.

Time is also an important factor. Let us now consider what each man who was able to see must have reasoned. Let us assume that case 5 is true, that the blind man has a black hat on his head. Both men who can see know that the blindman is wearing a black hat. If neither sees a black hat on the other man's head, momentarily neither can conclude anything further. But after a while, each would begin to realize that the other sighted man also does not see two black hats, at which time he would know his own hat had to be the white hat the other man saw because if it were black, the other man would know his own to be white.

Since neither sighted man spoke up to the King, we can conclude, as the blind man did, that he himself was not wearing a black hat, for he had been following the same reasoning.

Notice the use of inference in solving this problem. The silence of the men who could see was the key to our ability to infer what they must be thinking. Case 5 is thus proved impossible.

Computer Example

Consider the following programming problem that relies on inference. An excerpt from the 1978 Federal Income Tax Form is shown in Table 2.16.

Table 2.16   *Income Tax Table*

SCHEDULE X—Single Taxpayers

If the amount   Enter on Schedule
on Schedule TC,   TC, Part I:
Part I, line 3, is:

Not over $2,200 . . . . . .   –0–

Over—	But not over—		of the amount over—
$2,200	$2,700	14%	$2,200
$2,700	$3,200	$70 + 15%	$2,700
$3,200	$3,700	$145 + 16%	$3,200
$3,700	$4,200	$225 + 17%	$3,700
$4,200	$6,200	$310 + 19%	$4,200
.	.	.	.
.	.	.	.
.	.	.	.
$92,200	$102,200	$46,190 + 69%	$92,200
$102,200	—	$53,090 + 70%	$102,200

Assume that as a computer programmer for the government you are assigned the task of writing a program to process all income tax form inputs against this graduated tax table. A straightforward method would be to compose a conditional expression and successively test whether taxable income falls within each range specified. The program will then test for the occurrence of each range in the exact order given. For example, if the taxable income exceeds $102,200, the computer will make comparisons with all smaller amounts before discovering that fact. In this tax schedule, there are 25 entries.

Since there are millions of income tax forms to be processed, a built-in constraint is associated with this problem, the constraint that each be processed as efficiently as possible. We must therefore consider ordering the testing for income ranges so that a "match" is achieved within a reasonable amount of time. A common bit of information not explicitly stated in the problem is that the great majority of incomes in the United States fall below the $102,200 mark. This *implicit* problem given is the key to developing an efficient algorithm. We infer this information from past experience and knowledge. The relative inefficiency of the test for very high income thus does not worry us. A reasonable approach for calculating income tax appears in Example 2.12.

In some applications, however, we might wish to adopt a different strategy. For example, we might initially check to see if the taxable

Example 2.12 Income Tax Program

```
PROGRAM: TO SEARCH INCOME TAX TABLES

    input INCOME
    if INCOME less or equal 2200 then
        set INCOME_TAX  to    0
    else if INCOME less or equal 2700 then
        set INCOME_TAX  to    .14 * (INCOME - 2200)
    else if INCOME less or equal 3200 then
            set INCOME_TAX  to    70 + .15 * (INCOME - 2700)
    else if INCOME less or equal 3700 then
            set INCOME_TAX  to   145 + .16 * (INCOME - 3200)
    else if INCOME less or equal 4200 then
            set INCOME_TAX  to   225 + .17 * (INCOME - 3700)
    else if INCOME less or equal 6200 then
            set INCOME_TAX  to   310 + .19 * (INCOME - 4200)
    else if INCOME less or equal 8200 then
            set INCOME_TAX  to   690 + .21 * (INCOME - 6200)
    else if INCOME less or equal 10200 then
            set INCOME_TAX  to  1110 + .24 * (INCOME - 8200)
    else if INCOME less or equal 12200 then
            set INCOME_TAX  to  1590 + .25 * (INCOME - 10200)
    else if INCOME less or equal 14200 then
            set INCOME_TAX  to  2090 + .27 * (INCOME - 12200)
    else if INCOME less or equal 16200 then
            set INCOME_TAX  to  2630 + .29 * (INCOME - 14200)
    else if INCOME less or equal 18200 then
            set INCOME_TAX  to  3210 + .31 * (INCOME - 16200)
    else if INCOME less or equal 20200 then
            set INCOME_TAX  to  3830 + .34 * (INCOME - 18200)
    else if INCOME less or equal 22200 then
            set INCOME_TAX  to  4510 + .36 * (INCOME - 20200)
    else if INCOME less or equal 24200 then
            set INCOME_TAX  to  5230 + .38 * (INCOME - 22200)
    else if INCOME less or equal 28200 then
            set INCOME_TAX  to  5990 + .40 * (INCOME - 24200)
    else if INCOME less or equal 34200 then
            set INCOME_TAX  to  7590 + .45 * (INCOME - 28200)
    else if INCOME less or equal 40200 then
            set INCOME_TAX  to 10290 + .50 * (INCOME - 34200)
    else if INCOME less or equal 46200 then
            set INCOME_TAX  to 13290 + .55 * (INCOME - 40200)
    else if INCOME less or equal 52200 then
            set INCOME_TAX  to 16590 + .60 * (INCOME - 46200)
    else if INCOME less or equal 62200 then
            set INCOME_TAX  to 20190 + .62 * (INCOME - 52200)
    else if INCOME less or equal 72200 then
            set INCOME_TAX  to 26390 + .64 * (INCOME - 62200)
    else if INCOME less or equal 82200 then
            set INCOME_TAX  to 32790 + .66 * (INCOME - 72200)
    else if INCOME less or equal 92200 then
            set INCOME_TAX  to 39390 + .68 * (INCOME - 82200)
    else if INCOME less or equal 102200 then
            set INCOME_TAX  to 46190 + .69 * (INCOME - 92200)
    else    set INCOME_TAX  to 53090 + .70 * (INCOME - 102200)

    print 'INCOME TAX:', INCOME_TAX

end * program *
```

amount lies within the table, that is, is less or equal to $102,200, before searching the table. Such a strategy might be motivated by other inferred information.

In general, the sequence of table entries can affect the processing efficiency of table searching. If a table is very large, its structure greatly influences the search algorithm. Entries may be in ascending order, descending order, or no particular order. Because a table search starts at the top, the proper entry will always be found eventually provided it exists.

Sometimes table entries can be sequenced to maximize the probability of a quick "hit." For example, if we are able to infer that 70 percent of the hits will be made against a specific set of entries, these entries should be placed at the beginning of the table. When hits are more evenly distributed, it is not possible to structure a large table in this way. In some cases, a table search may not even be practical.

Review of the Basic Idea

Inference is one of the first tools to be applied in the problem definition step. With it, we draw additional information from the problem givens. The squeezing of liquid from a ripe orange might make an appropriate metaphor. The juices in the orange represent all the knowledge we have concerning a problem, both that explicitly given and any problem relationships stored in our memory. The additional information generated by the "squeeze" of inference may bring forth the "core" essence of the problem and be the critical factor in the problem solution.

In studying the method of working backwards, we focused on the goal and considered it as the starting point for the problem-solving process. In a sense, we used inference to work backwards, drawing inferences concerning the goal that would indicate the operations and givens which preceded it.

PRESCRIPTION 9

Every Noble Work Is at First Impossible

Sometimes it is wise to solve a problem by breaking it up into parts, which may be easier to handle than an unwieldy whole. The mind can become saturated if it attempts to process too many inputs at once.

This method is called "subgoaling." It consists of replacing a single problem with two or more simpler problems. Since it is particularly advantageous for attacking nontrivial problems requiring a sequence of actions, it is ideal for programming problems, which involve an orderly sequence of computer instructions.

In subgoaling, each subgoal is a subproblem, or smaller problem within a larger problem, and consists of its own set of givens, goal, and operations, as shown in Fig. 2.14.

The value of subgoaling is that it allows the problem solver to concentrate on each subproblem independently of other subgoals, and even the larger problem, and thus to avoid dealing with an avalanche of inputs.

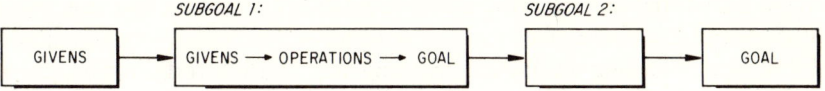

Fig. 2.14 *Subgoaling approach to problem solving*

The first task is to define a set of problem subgoals that are unique to the larger problem.

Non-Computer Example

As an example of subgoaling, consider the problem illustrated in Fig. 2.15. Nine men and two boys want to cross a river on an inflatable raft that will carry only one man or two boys at one time. How many times must the boat cross the river to accomplish this goal? (A round trip equals two crossings.)

A convenient subgoal for this problem would be to transport *one* man across the river and return the boat to the starting side, for it is clear that if we can do this once, we can do it eight more times and thus solve the larger problem.

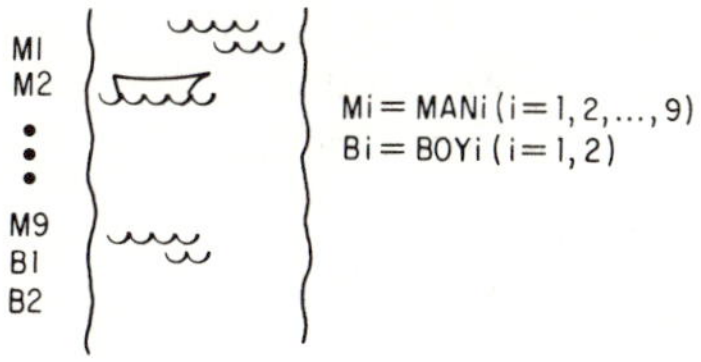

$$M_i = MAN_i\ (i = 1, 2, ..., 9)$$
$$B_i = BOY_i\ (i = 1, 2)$$

Fig. 2.15 *Another river crossing problem*

One way to solve this subgoal, as shown in Fig. 2.16, is for the two boys to cross the river in the boat first. Then one boy takes the boat back to the original side. A man then takes the boat across the river, and the second boy takes the boat back to the original side. It thus takes four crossings to transport one man across the river and return the boat to the original side. Both boys are now on the original side, one man is on the other side, and there are eight men left.

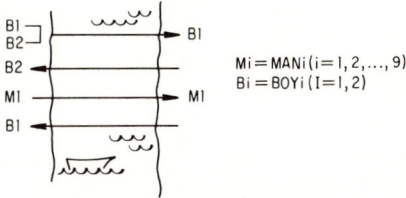

$$M_i = MAN_i (i = 1, 2, ..., 9)$$
$$B_i = BOY_i (I = 1, 2)$$

Fig. 2.16 *Subgoal solution*

To transport all nine men across the river, it will thus require 9 × 4, or 36 one-way crossings. Eventually, both boys will be on the original side with the boat, and one additional crossing will be required for them to get to the other side. Thus a total of 37 one-way trips will be required to solve the problem.

This example illustrates the power of subgoaling to simplify the problem-solving process. Only one subgoal was involved here—to transport one man across a river—but other problems may have multiple subgoals. An advantage of subgoaling in this example is that we can use the same subgoal logic over and over.

The analogy of the subgoal in programming is the module, a distinct unit of logic or subproblem. Two types of modules are known as *functions* and *subroutines*.

A function is an operator that performs a transformation on its arguments and produces a single resultant value. For example, if we want to find the square root of a number, we might write:

$$x = SQRT\ (100)$$

SQRT is a system library type of function supplied by many programming languages, and the variable x is automatically assigned the square root of 100, or 10. The argument in this case, 100, is not modified. When more than one argument is used, they are separated by commas.

Any reference to a function name in a program will cause a resultant value to be produced, including a function name in an expression. For example,

$$x = 10 + 5 * SQRT\ (4)$$

will cause the function SQRT to evaluate expression (4) and assign $10 + 5 * 2$, or 20, to the variable x.

Sometimes it is necessary for a user program to provide its own user function to perform specific calculations. For example, a program may require the average of three numbers at different places in the program. The use of a user-designed function for this purpose is illustrated in Fig. 2.17.

All references to the function name AVERAGE will cause the associated logic defined by the function to operate on the argument list that is

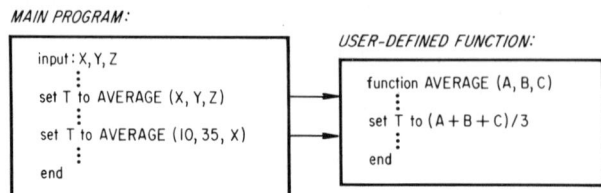

Fig. 2.17 *Example of a user function*

passed from the main program to the subroutine. In the above example, the variable T will be assigned to the average of variables X, Y, and Z and at a later time in the program will be assigned to the average of 10, 35, and X. A function is thus a convenient computation mechanism; conceptually, it is a subgoal.

A subroutine is very similar to a function, although it is not assigned a single value. It is more flexible and does not necessarily return a resultant value, but it may perform calculations and print the results. Also, a subroutine may or may not be defined with arguments, whereas a function usually requires at least one argument.

The vehicle for referencing a subroutine is the CALL statement. As an example of the use of a subroutine, consider the main program in Fig. 2.18; this calculates the average of five arguments and prints the result at various points in the program.

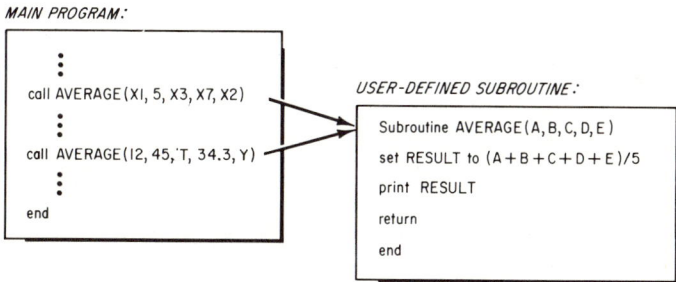

Fig. 2.18 *Example of a subroutine*

The number of arguments in the CALL statement must agree with the number of arguments in the subroutine definition. In this example, the user-defined subroutine performs all the defined calculations and outputs and then passes control back to the main program, one statement after the original CALL statement. Again, the subroutine is conceptually a subgoal of a larger problem.

Another application of subroutines involves the payroll problem posed earlier, concerning the processing of a payroll for a group of employees. A program problem solution with a subroutine approach is shown in Example 2.13.

Example 2.13 *Payroll Program (Subroutine Approach)*

```
PROGRAM: TO PROCESS PAYROLL

   for CURRENT_EMPLOYEE = 1 to NUM_EMPLOYEES do the following

      input EMPLOYEE_ID, HOURS_WORKED, PAY_RATE, TOTAL_SECURITY
```

```
                                                 . . . . . . . . . . . . .
     call INCOME_TAX(...) -------------------->  . subroutine .
                                                 . . . . . . . . . . . . .

                                                 . . . . . . . . . . . . .
     call SOCIAL_SECURITY_AMT(...) ----------->  . subroutine .
                                                 . . . . . . . . . . . . .

                                                 . . . . . . . . . . . . .
     call DEDUCTIONS(...) -------------------->  . subroutine .
                                                 . . . . . . . . . . . . .

                                                 . . . . . . . . . . . . .
     call NET_PAY(...) ----------------------->  . subroutine .
                                                 . . . . . . . . . . . . .

                                                 . . . . . . . . . . . . .
     call PRINT_PAY_STUB(...) ---------------->  . subroutine .
                                                 . . . . . . . . . . . . .

                                                 . . . . . . . . . . . . .
     call PRINT_SUMMARY_REPORT(...) ---------->  . subroutine .
                                                 . . . . . . . . . . . . .
end * program *
```

3

Advanced Problem-Solving Prescriptions

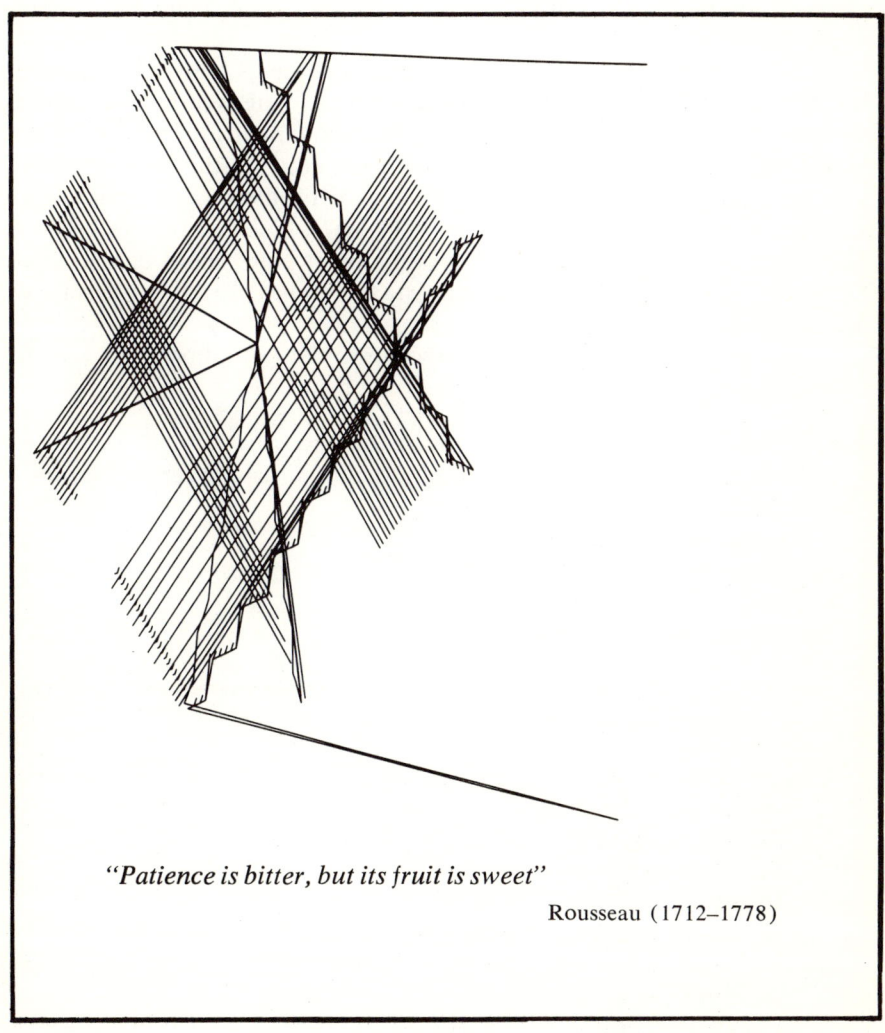

"Patience is bitter, but its fruit is sweet"

Rousseau (1712–1778)

PRESCRIPTION 1

Zeal Without Knowledge Is the Sister of Folly

The first question to be asked in problem solving is, "Is there a problem?" The answer may appear obvious, but, in fact, this question is usually not addressed in problem-solving situations. It is often merely assumed that there is a problem without analyzing the circumstances surrounding the specific situation.

The distinction that must be made is between real problems and mere annoyances. A problem can be considered a situation that is perplexing, and attempting to solve it is accepted as a meaningful endeavor by most rational individuals. An annoyance, on the other hand, is no more than a minor irritation.

It is important to distinguish between problems and annoyances so that we may concentrate our efforts towards solving worthwhile difficulties rather than petty distractions. When annoyances are treated as real problems, wasted efforts and additional problems are the usual results.

Ironically, the distinction between real problems and annoyances has been clouded by the computer revolution, which has relieved man of the mental burden associated with mountains of voluminous calculations. The very availability of the computer leads all too easily to the assumption that it is the panacea for every ill.

The following list summarizes those problem situations that warrant a possible computer solution:

1. Lengthy tasks requiring a great deal of time
2. Recurring tasks in which a program can be used over and over
3. High level of accuracy required
4. High level of reliability required
5. Storage of a great deal of data for long periods of time
6. Output of a great deal of data
7. Decision making on an elementary level

Some general areas in which the computer has been used effectively to solve real-world problems are the following:

1. General Business
    Accounts receivable
    Accounts payable
    Personnel accounting
    Payroll
    Inventory control
2. Banking
    Account reconciliation
    Installment loan accounting

56

      Interest calculating
      Demand deposit accounting
      Savings
      Trust services
3. Education
      Attendance and grade reports
      Computer-assisted instruction
      Research analysis
4. Government
      Income-tax return verification
      Motor vehicle registration
      Budget analysis
      Tax billing
      Property rolls
5. Other Areas
      Law enforcement
      Military
      Sports
      Transportation
      Broadcasting
      Real estate
      Business forecasting
      Medicine

In summary, it is important to utilize the computer to solve real-world problems, not annoyances. Applications not warranting computer solutions can usually be handled by manual methods, or best of all, ignored.

PRESCRIPTION 2

Let Your Imagination Run Wild

When we experience mental blocks in problem solving, what we need to do is find a way to neutralize them effectively. To do so, we should try a new approach in which our logical path will be completely different from those of previous attempts. Any new path we take will provide a real alternative to the one that created the mental block.

In general, mental blocks arise as a result of stress. The ingredient needed to eliminate them is developing a greater imagination. Imagination is a basic tool for changing directions in our thinking process, for freeing us from limitations in our reasoning ability. It provides us with new directions.

Often we may exclude a workable problem-solving method by a quick analysis before we have had a chance to see if it offers any possibilities. In order to illustrate this failing, consider the following problem, which instructs us to connect all nine dots illustrated in Fig. 3.1 with only four straight lines. We may not let our pencil leave the paper, and we may use only straight lines.

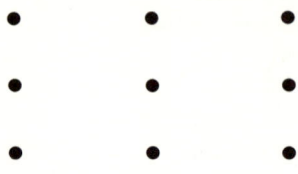

Fig. 3.1   Nine dots problem

Invariably, most people attempting to solve this problem fall into the same trap. That is, they add a constraint that is not part of the problem, namely, keeping all lines within the boundary of the nine dots. Notice that the stated goal does not include this constraint. Why then do people inject it into the problem? One explanation is that our minds want the problem to involve only the given dimensions, as if anything else might be cheating. If this special constraint is placed on the problem, however, it becomes unsolvable. The imaginative solution to the problem is shown in Fig. 3.2.

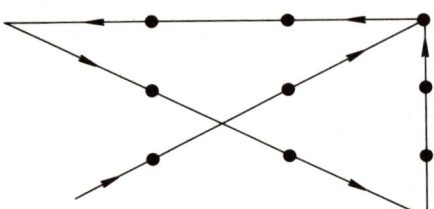

Fig. 3.2   Solution to nine dots problem

Notice that the solution extends well outside the boundary of the nine dots. If we attempted to keep the lines within the boundary, the inevitable result would be a mental block, and we would begin to retrace the same set of diagrams over and over again.

Computer Examples

To illustrate the value of imagination further, consider the following mathematical problem. Pi, the sixteenth letter of the Greek alphabet, designates the mathematical ratio of the circumference of a circle to its diameter. This important concept has been known for centuries. For example, the Egyptians used it in their development of the pyramids. With the use of the computer, the problem of calculating pi is easily solved.

The following clever algorithm illustrates how imagination was brought to bear in solving this problem. Suppose we inscribe a circle of unit radius into a square whose sides are two units, as shown in Fig. 3.3. Knowing that the area of a circle (A) with unit radius (r) is pi (π)—since $A = \pi r^2$ and $r^2 = 1$—and that the area of the square in which the circle is circumscribed is 4 (or 2 $\times$ 2), we can find the value of pi by dividing the area of the circle by the area of the square and multiplying by 4, that is, $(\pi/4) \times 4 = \pi$. The same argument holds if we apply this procedure to the quarter of a circle which is in the first quadrant of the X-Y plane and the unit square in which the quarter of the circle is inscribed.

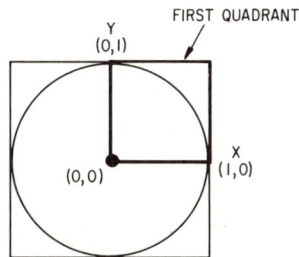

Fig. 3.3   *Unit circle inscribed in square*

By generating two random numbers between 0 and 1 corresponding to the values of the X, Y coordinates of a point within the unit square and knowing that the formula of the circle is $X^2 + Y^2 = 1$, we can count the number of points falling within the quarter of the circle. For instance, if we generate a total of 10,000 points in the unit square, the value of pi would be given by the algebraic equation:

$$\text{Pi} = \frac{4 \times (\text{Number of points falling within the quarter circle})}{10,000}$$

We now have a simple method for determining the value of pi. The program for this algorithm is illustrated in Example 3.1.

Another problem involving exercise of the imagination concerns the shuffling of a deck of ordinary playing cards. At first glance, this problem might appear to be quite remote from a computer solution, but a little play of fancy can solve it.

A deck of playing cards consists of 52 cards in four suits, the cards ranging from a deuce (2) to an ace. Shuffling the deck of cards involves arranging the 52 cards in a random order so that adjacent cards have no apparent relationship in terms of card type or suit. With some imagination,

Example 3.1 *Approximation of Pi*

```
PROGRAM: TO COMPUTE PI

Constant Definitions

   TOTAL_TRIALS = 10000
   seed         = 0

   set NUM_POINTS  to 0
   for TRIAL_NUM = 1 to TOTAL_TRIALS do the following

      set X  to  random number (0,1)
      set Y  to  random number (0,1)
                     2    2
      set POINT to x  +  y
      if  POINT less or equal 1 then

         add 1 to NUM_POINTS

   set PI_APPROXIMATION  to   4 * (NUM_POINTS / TOTAL_TRIALS)
   print 'PI:',PI_APPROXIMATION

end  * program *
```

Table 3.1 *Card Type and Computer Table Position*

Card Type	Card Value	Card Type	Card Value
Two of spades	1	Eight of hearts	27
Two of clubs	2	Eight of diamonds	28
Two of hearts	3	Nine of spades	29
Two of diamonds	4	Nine of clubs	30
Three of spades	5	Nine of hearts	31
Three of clubs	6	Nine of diamonds	32
Three of hearts	7	Ten of spades	33
Three of diamonds	8	Ten of clubs	34
Four of spades	9	Ten of hearts	35
Four of clubs	10	Ten of diamonds	36
Four of hearts	11	Jack of spades	37
Four of diamonds	12	Jack of clubs	38
Five of spades	13	Jack of hearts	39
Five of clubs	14	Jack of diamonds	40
Five of hearts	15	Queen of spades	41
Five of diamonds	16	Queen of clubs	42
Six of spades	17	Queen of hearts	43
Six of clubs	18	Queen of diamonds	44
Six of hearts	19	King of spades	45
Six of diamonds	20	King of clubs	46
Seven of spades	21	King of hearts	47
Seven of clubs	22	King of diamonds	48
Seven of hearts	23	Ace of spades	49
Seven of diamonds	24	Ace of clubs	50
Eight of spades	25	Ace of hearts	51
Eight of clubs	26	Ace of diamonds	52

however, shuffling cards can be considered to be the same problem as arranging a table of 52 computer data items in random order.

To implement this comparison, we must associate each card in the deck with the position of a data item in a computer memory. Table 3.1 illustrates one way that this can be accomplished although the order is arbitrary.

Next, we must be able to represent a computer table that is 52 positions in length. This can be done easily with a collection of two or more adjacent computer memory cells that are associated with a single symbolic name. Each memory cell is called a *table entry*. For example, suppose we create a table called CARD—DECK to represent the cards in our 52-card deck. The subscripted variable CARD—DECK (1) can be used to reference the first entry of the table; CARD—DECK (2), the second; and CARD—DECK (52), the last entry. The integer enclosed in parenthesis is called the *table subscript* and may also contain an integer variable.

We will use a total of two subscripted tables in the program to shuffle each of the 52 elements. The first table, called DUP—TABLE (CARD), will ensure that none of the 52 random numbers generated is duplicated. DUP—TABLE(I) will have the value 1 if the number represented by the subscript I has already been generated. If DUP—TABLE (I) is assigned the value 0, then the number represented by its subscript has not yet been generated.

The array called CARD—DECK will contain the 52 numbers in random order and will be equivalent to the shuffled deck of cards when processing is completed.

The program to shuffle and print cards is shown in Example 3.2.

Example 3.2 *Program to Shuffle Deck of Playing Cards*

```
PROGRAM: TO SHUFFLE A DECK OF PLAYING CARDS

Constant Definitions

    RANK_TABLE(1)  =  '2'
    RANK_TABLE(2)  =  '3'
    RANK_TABLE(3)  =  '4'
    RANK_TABLE(4)  =  '5'
    RANK_TABLE(5)  =  '6'
    RANK_TABLE(6)  =  '7'
    RANK_TABLE(7)  =  '8'
    RANK_TABLE(8)  =  '9'
    RANK_TABLE(9)  =  'T'   ** TEN **
    RANK_TABLE(10) =  'J'   ** JACK **
    RANK_TABLE(11) =  'Q'   ** QUEEN **
    RANK_TABLE(12) =  'K'   ** KING **
    RANK_TABLE(13) =  'A'   ** ACE **
    NUM_CARDS_DECK =  52
    NUM_RANKS      =  13
    seed           =  0
```

```
* Clear Duplication Table *
for CARD_COUNT = 1 to NUM_CARDS_DECK do the following

   set DUP_TABLE(CARD_COUNT)  to  0

* place random card in deck *
for CARD_COUNT = 1 to NUM_CARDS_DECK do the following

   repeat until DUP_TABLE(CARD) not equal to 1

      set CARD  to integer portion: (RANDOM x 51) + 1

   set DUP_TABLE(CARD) to 1
   set CARD_DECK(CARD_COUNT)  to  CARD

* OUTPUT SHUFFLED CARD DECK *
for CARD_COUNT = 1 to NUM_CARDS_DECK do the following

   set SUIT  to remainder of (CARD_DECK(CARD_COUNT) / 4)
   set RANK  to  integer portion of (CARD_DECK(CARD_COUNT) / 4)

   if SUIT = 1 then
      print RANK_TABLE(RANK + 1),' OF SPADES'

   else
      if SUIT = 2 then
         print RANK_TABLE(RANK + 1),'OF CLUBS'

   else
      if SUIT = 3 then
         print RANK_TABLE(RANK + 1),'OF HEARTS'

   else
      print RANK_TABLE(RANK),'OF DIAMONDS'

end  * program *
```

First of all, the entries of DUP_TABLE are initialized to zero. Then random integers ranging from 1 to 52 are generated by the statement

> set CARD to integer portion: $(RANDOM \times 51) + 1$

In this expression, a random number is first generated between 0 and 1 by the user-defined function RANDOM. This number is then multiplied by 51, and the integer portion of the product is computed. Finally, 1 is added to this result to obtain a random number between 1 and 52. This formula can be checked by generating lower and upper bounds for the RANDOM function—0 and 1, respectively. For the former case, the final result, it will be 1, and for the latter case, the result, CARD, it will be 52. The statement

> repeat until DUP_TABLE (CARD) not equal to 1

is used to check if the random number generated is duplicated. If no duplicate is found, the statement set DUP__TABLE (CARD) to 1

records the fact that the new number, CARD, has been generated. If a duplicate random number is found, another random number is generated until a duplicate number is not found. The statement

set CARD__DECK (CARD__COUNT) to CARD

is used to save the random number between 1 and 52 and place it in the appropriate CARD__DECK position. This is equivalent to placing a card randomly in the deck. The above process continues until the table CARD__DECK contains random numbers between 1 and 52, that is, when the card deck is "shuffled." The card deck is then symbolically output by rank and suit.

In summary, our imagination is an important part of problem solving. Its creativity enables us to form ideas free of ordinary constraints. In such an atmosphere, the mind begins to release itself from the restrictions inhibiting it from visualizing new approaches.

If imagination is so important to our ability to solve problems, perhaps we should try to find ways in which we could improve it! Normally, our imagination in problem solving depends upon the experience we have had with problem-solving methods. The more methods we have been exposed to, the easier it becomes for us to think of different ways in which we can attack a problem. Hence our imagination can truly be improved by experience.

PRESCRIPTION 3

Extinguish Fire with a Brainstorm

A primitive thinking tool to spur use of the imagination in problem solving is called "brainstorming." It consists of learning to think of all ways to do a particular thing without first evaluating the worthwhileness of any of them. Brainstorming is an effective way of letting our imaginations run wild. It is an exercise in considering all possible ideas regardless of how ridiculous they may seem. Often those that initially appear to have no merit become the very ones that lead us to success. An analogy, in a sense, is constructively "grasping for straws."

Non-Computer Example

To show how brainstorming can be a very effective means of problem solving, we offer the following senario.

Your spaceship has just crash-landed on the lighted side of the moon. You were scheduled to rendezvous with a mothership already on the moon but 200 miles away. The rough landing has wrecked your ship and destroyed all the equipment on board except for the 15 items listed in

Table 3.2   *List of Survival Items*

Box of matches
Food concentrate
Fifty feet of nylon rope
Parachute silk
Solar-powered portable heating unit
Two .45-caliber pistols
One case of dehydrated milk
Two 100-pound tanks of oxygen
Stellar map (of the moon's constellation)
Self-inflating life raft
Magnetic compass
Five gallons of water
Signal flares
First-aid kit containing injection needles
Solar-powered FM receiver-transmitter

Table 3.2. Since your crew's survival depends on reaching the mothership, you must choose the most critical of the items that are available for the 200-mile trip.

The problem is to rank the 15 items in terms of their importance for survival. To do so, place number 1 by the most important item, number 2 by the second most important, and so on through number 15, the least important. Incidentally, four to seven persons should attempt to solve this

Table 3.3   *Solution to Moon Rendezvous Problem*

Item	*NASA's reasoning*	*NASA's rank*
Box of matches	No oxygen on moon to sustain flame	15
Food concentrate	Efficient means of supplying energy	4
Fifty feet of nylon rope	Useful in scaling cliffs, tying injured together	6
Parachute silk	Protection from sun's rays	8
Solar-powered portable heating unit	Not needed unless on dark side	13
Two .45-caliber pistols	Possible means of self-propulsion	11
One case of dehydrated milk	Bulkier duplication of food concentrate	12
Two 100-pound tanks of oxygen	Most pressing survival need	1
Stellar map	Primary means of navigation	3

Item	NASA's reasoning	NASA's rank
Self-inflating life raft	CO_2 bottle in military raft may be used for propulsion	9
Magnetic compass	Magnetic field on moon is not polarized	14
Five gallons of water	Replacement for tremendous liquid loss on lighted side	2
Signal flares	Distress signal when mother ship is sighted	10
First-aid kit containing injection needles	Needles for vitamins, medicines, etc.	7
Solar-powered FM receiver-transmitter	For communication with mother ship; but FM requires line-of-site transmission and short ranges	5

SCORING:

0–25	Excellent
26–32	Good
33–45	Average
46–55	Fair
56–70	Poor
71–112	Very Poor (suggests possible faking or use of earth-bound logic)

problem. The problem should be solved individually first, without knowing the other's answers, and then as a group. When the problem is solved as a group, individual solutions should be shared until a consensus is reached—one ranking for each of the 15 items that best satisfies *all* group members.

The grading mechanism for evaluating the individual or group answers is the absolute difference between the item answers and known solutions, that is, the lower the score, the better the score. The known solutions, supplied by NASA, are shown in Table 3.3.

The spaceship rendezvous problem is an excellent exercise in brainstorming, for it is very complex and requires input from various technical disciplines, as might be obtainable from a group. As a matter of fact, this is the essence of brainstorming.

Figure 3.4 illustrates the typical results achieved when attempts are made to solve the problem individually.

In this sample, a group of 20 members was broken up into three groups, with resultant scores shown in Table 3.4.

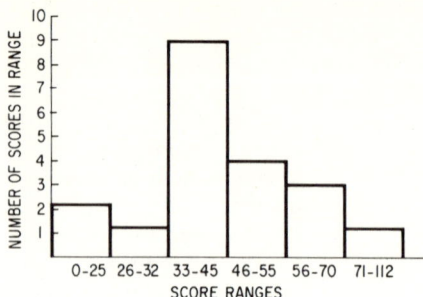

Fig. 3.4 *Typical individual responses to moon problem*

Table 3.4 *Summary of Sub-Group Results*

Group No.	1	2	3
Sub-group size	5	6	9
Sub-group score	14	18	38
Rating	E	E	A

Where: E = Excellent
A = Average

It is interesting to note that Group 3, which was composed of 9 members, did not score as well as the smaller groups with 5 and 6 members, respectively, which suggests that there exists an optimal number of individuals for group brainstorming. The performance of Groups 1 and 2 was greatly superior to that of individuals, where the average was 43.95.

In summary, brainstorming can be a very useful problem-solving method for complex problems that require the knowledge of a variety of individuals. The flow of ideas of a group may enhance the creative process and provide more avenues of approach.

Computer Examples

The following discussion illustrates how brainstorming can also be useful in programming. Once a program has been designed and presumably solved, it is important that the design be carefully reviewed before entering the coding stage. In a process similar to proofreading in the publishing industry, programmers and designers should proofread their designs.

A design "walk-through" consists of a group of reviewers who carefully analyze the program design. The group consists of the program designer, or problem solver, plus a number of independent, experienced programmer reviewers. The designer carefully "walks" the group through the design. This process is meant to uncover program design errors, omissions, inconsistencies, and contradictions. However, bugs are not eliminated at this time, only recorded; the designer makes the corrections later. The design walk-through is thus a process in which a program design is checked for correctness; the group merely attempts to prove or disprove it suitability.

Another application of brainstorming in programming is code walk-throughs. A code walk-through is the same type of process as the one just described and consists of a group of programmers carefully reading and analyzing actual program listings. The virtue of the code walk-through is its ability to identify significant numbers of coding and logic errors.

Finally, brainstorming can be used in the program design phase. Often a program or programming system is so complex that it initially requires the combined ideas of a group. As with design and code walk-throughs, the group works as a whole and considers all possible ideas, regardless of how ridiculous they may seem.

PRESCRIPTION 4

The Shell Must Break Before The Bird Can Fly

Problem solutions have an aspect that is sometimes ignored: their practicality. To emphasize how important practicality is, consider the mechanical contraptions of the cartoonist Rube Goldberg. While their methods may actually solve the problem at hand, they can hardly be considered practical!

Computer Example

The following programming problem illustrates the point we have just been making. You are given a table of positive numbers, of length NUM_ENTRIES, that you must sort into ascending order. This table will be called UNSORTED_TABLE. A brute-force method for solving this problem would be to create a target table first that will ultimately contain the sorted data from UNSORTED_TABLE. This table will be called SORTED_TABLE, as shown in Fig. 3.5. The table size, NUM_ENTRIES, will be called "N."

To sort the table, a pointer is set to the first position in SORTED_TABLE, and a "very large number" is placed in this position. UNSORTED_TABLE is then searched for a number less than the current value in the first position of SORTED_TABLE. If such a number is

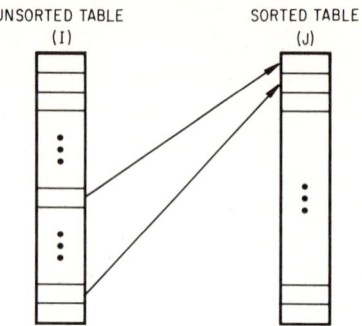

Fig. 3.5   *Sorting a table*

found, it is stored in the first position of SORTED_TABLE. After the complete UNSORTED_TABLE has been searched, the first position of SORTED_TABLE will contain the smallest number from UNSORTED_TABLE. At this point, the smallest table entry in UNSORTED_TABLE is set to a minus one so that it will not be processed again (only positive numbers are processed).

Next, the SORTED_TABLE pointer is incremented by one, and the above process is repeated. When SORTED_TABLE is completely filled, it will contain a sorted list of ascending data items from UNSORTED _TABLE, and UNSORTED_TABLE will contain all minus one's. A program reflecting this brute-force approach is shown in Example 3.3.

Example 3.3   *Sort Data Table (Impractical)*

```
PROGRAM: TO SORT DATA TABLE (impractical)

Constant Definitions

  NUM_ENTRIES = 50

  for SORT_CURRENT_ENTRY = 1 to NUM_ENTRIES do the following
     input UNSORTED_TABLE(SORT_CURRENT_ENTRY)

  for SORT_CURRENT_ENTRY = 1 to NUM_ENTRIES do the following

     set SORTED_TABLE(SORT_CURRENT_ENTRY)  to  1000000000

     * Search UNSORTED_TABLE for smallest value *
     for UNSORTED_CURRENT_ENTRY =1 to NUM_ENTRIES do the following

        if UNSORT_TABLE(UNSORTED_CURRENT_ENTRY) greater or equal 0 then

           if UNSORTED_TABLE(UNSORTED_CURRENT_ENTRY) less than
              SORTED_TABLE(SORT_CURRENT_ENTRY) then

              set SORTED_TABLE(SORT_CURRENT_ENTRY)  to
                 UNSORTED_TABLE(UNSORTED_CURRENT_ENTRY)
```

```
          set  LARGEST_ENTRY  to  UNSORTED_CURRENT_ENTRY

     set  UNSORTED_TABLE(LARGEST_ENTRY)    to   -1

end   * program *
```

With a little thought, this method is seen to be not very practical, and for the following reasons:

1. Method requires N^2 comparisons.
2. Only a table of positive numbers can be sorted.
3. Two tables are required.
4. Sorted result is in SORTED_TABLE.
5. UNSORTED_TABLE is destroyed.
6. There is no way to be assured that the "very large number" is large enough.

A more practical method for sorting a table into ascending order is called the "selection sort." Instead of an additional table being needed to process the sort, only one additional temporary variable is required.

As shown in Fig. 3.6, the value of the first position in SORTED_TABLE is compared with the value in the next position of SORTED_TABLE. If the value of the first position is greater than the value of the second position, they are interchanged as follows:

1. The value of the first position of SORTED_TABLE is stored in the temporary variable.
2. The value of the second position of SORTED_TABLE is stored in the first position.
3. The value of the first position is moved from the temporary storage area to the second position of SORTED_TABLE.

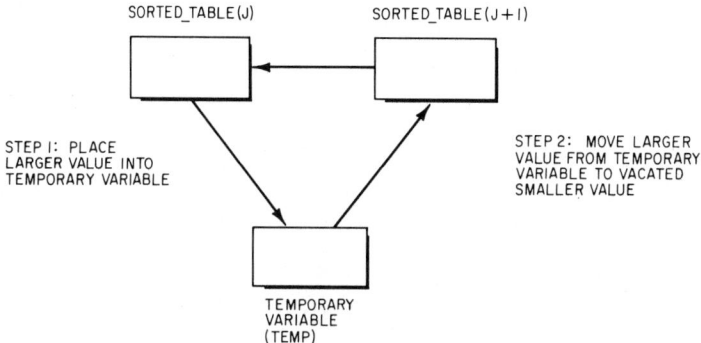

Fig. 3.6 *Selection sort process*

The first pass through SORTED—TABLE is completed when positions 2, 3, . . . , N have been compared with the value of the first position of SORTED—TABLE. At this point, the first position of SORTED—TABLE contains the smallest value and is left alone. Next, the second position of SORTED—TABLE is compared with each value in positions 3, 4, . . . , N during the second pass through the SORTED—TABLE.

With each successive pass through SORTED—TABLE, one less comparison is required; after N – 1 passes, the SORTED—TABLE is in numerically ascending order. A program that uses this technique is shown in Example 3.4.

Example 3.4 *Sort Data Table (Selection Sort)*

```
PROGRAM: TO SORT DATA TABLE (selection sort)

Constant Definitions

  NUM_ENTRIES = 50

  for CURRENT_ENTRY = 1 to  NUM_ENTRIES  do the following
      input SORTED_TABLE(CURRENT_ENTRY)

  for CURRENT_ENTRY = 1 to (NUM_ENTRIES - 1) do the following

     for NEXT_ENTRY = CURRENT_ENTRY to (NUM_ENTRIES - 1)
                  do the following

       if SORTED_TABLE(NEXT_ENTRY + 1) less than
          SORTED_TABLE(CURRENT_ENTRY) then

          set  TEMP  to  SORTED_TABLE(CURRENT_ENTRY)
          set  SORTED_TABLE(CURRENT_ENTRY)  to
               SORTED_TABLE(NEXT_ENTRY + 1)

          set  SORTED_TABLE(NEXT_ENTRY + 1)  to TEMP

end  * program *
```

The selection sort is more practical than the first method for the following reasons:

1. Only $.5N(N - 1)$ comparisons are required, not N^2.
2. Positive or negative numbers can be sorted.
3. Only a temporary variable is required.
4. Sorted results are maintained in SORTED—TABLE.

To emphasize the superior efficiency of the selection sort, as compared to the previous method, Table 3.5 shows the number of sort comparisons required in each case, as a function of the table size, N.

In addition to the importance of the practicality of a solution, a program must also be usable in terms of the user. Often a program is so

Table 3.5 Number of Comparisons Required

N	N^2	$.5N(N-1)$
5	25	10
15	225	105
25	625	300
50	2500	1225
100	10000	4950

complicated that it is very difficult to implement. For example, user input interfaces may be so cumbersome that the user is at a loss. More time is spent trying to understand how to use the program than applying it to solving a problem. In fact, a whole set of new problems are created if a program is not readily usable.

It is therefore essential to realize that while a program may have to be complex, the user of the program must be able to implement the solution easily.

PRESCRIPTION 5

Know Thyself

One of the most important psychological considerations in program problem situations is the ego-drive—the constant attempt to maintain or enhance one's feeling of self-worth. The saying "When a man tries himself, the verdict is in his favor" illustrates the point. The ego self-image varies greatly from time to time—contracting, expanding, and marking time as it interacts with the environment. This is not to deny the fact that the ego is a stable and recurring phenomenon.

Because of its shifts of mood, the ego can be considered a double-edged sword in that it can either aid us in problem solving or distract us. For example, its deflation due to a momentary lack of confidence may inhibit the problem-solving process. When it is threatened, its natural reaction is self-defense and cover-up. As a consequence, energies are often expended on restructuring self-worth rather than on problem solving.

On the other hand, the over-inflated ego can also interfere with effective problem solving. For example, all sorts of mechanisms may again be employed to protect the ego when it feels forced to defend its own solution approach. The trick to effective problem solving is to produce a happy medium between the two ego extremes. Needless to say, however, a prerequisite to effective problem solving is a positive sense of self-worth.

As discussed earlier, a technique in programming that attempts to foster this happy balance is the "walk-through," which promotes an atmosphere of problem solving that does not altogether discourage the self-image. In this environment, programmers are reinforced and rewarded for sharing ideas.

This prescription basically says that you should use your ego to your advantage but be aware of its power during the problem-solving situation to be productive or destructive.

PRESCRIPTION 6

Incubate When Gears Get Stuck

Often our attempts to solve a problem reach a point where our thinking seems to go in circles, our analysis apparently leading nowhere. This phenomenon is called a *mental block*. Mental blocks are not only very frustrating, they can totally impede problem solving. As an illustration, let us review the basic problem-solving process.

The process of solving a problem is to operate on the information provided—the givens—in a way that will produce the solution sought—the goal. Between the givens and the goal, a set of operations occur that change the status of the problem. Figure 3.7 illustrates the mental blocks that may interfere with this process.

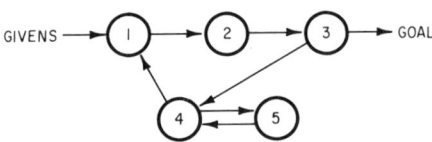

Fig. 3.7   *Going around in circles*

It is obvious from the figures that the logical sequence is from the givens, through states 1, 2, and 3, to the goal. It is only too possible, however, that we will fail to recognize this path immediately and enter into a circular pattern instead. Somehow the relationships that will take us from state 3 to the goal elude us. We may proceed through the cycle any number of times before discovering that we have entered a loop that seems to have no exit. It is as if our brain were "locked" into position, unable to change direction.

Imagination can be a very effective remedy for mental blocks, but a problem may be so complex that even the imagination may be stymied. The present prescription emphasizes the importance of getting completely away from certain problems and letting them incubate on their own. When the mind is free of the need to solve a problem, the reason for the block may well make itself known, permitting us to find a new approach.

The problem should be put aside for minutes, hours, days (perhaps even months or years), and we should work on something else, go fishing, or refresh ourselves in sleep. That skilled problem solvers religiously believe in incubation is evidenced by the close availability of beds for scientists working on the A-bomb during World War II. Another example of the benefits of incubation may be seen in the realm of artistic creativity. It took Beethoven four years to write the Fifth Symphony. In programming, the power of incubation is well known. Many the programming problem that has been solved by a programmer while driving a car.

Psychologists don't really understand how or why incubation works, but there are at least three theories that attempt to explain it. The first stresses the release from pure fatigue that sleep and relaxation provide, thereby restoring the mind to a more functional level. The second stresses the role of memory. The mind becomes so preoccupied with its recollection of its many failures, so intimidated by its circular behavior, that the creativity process is thoroughly stunted. With the passage of time, memories of past mistakes fade away, and the mind is free to engage in fresh approaches. The third theory stresses unconscious forces. Incubation allows the mind to work unconsciously on the problem after the conscious mind has set it aside.

PRESCRIPTION 7

A Pound of Analogy Equals a Ton of Sweat

Another powerful primitive thinking tool for problem solving is called "analogy." An analogy is a revelatory correspondence. To make an analogy is to explain something by its resemblance to something else.

The metaphor, a figure of speech in which one thing is likened to another, is a form of analogy. It is an implied comparison, in which a word or phrase ordinarily used to signify one thing is applied to another. For example, a farmer who has never seen a ship navigate in the ocean might be helped to imagine it if told that the ship "plows" the water as it sails through.

A form of analogy more relevant to programming posits the solution of a problem as a special case of a known general problem or involves the solution of a general problem from a known special case.

The application of analogy to problem solving can establish equivalence as well as similarity. "Equivalent problems" are exactly the same, whereas "similar problems" are alike yet not the same in every respect.

Non-Computer Example

Consider the following geometry problem as an illustration of the technique of analogy. Given the following problem information, you are to prove that the sum of the angles of any triangle equals 180 degrees:

1. A straight line equals an angle of 180 degrees
2. All right angles equal 90 degrees
3. If two parallel lines are cut by a transversal, the corresponding interior angles are equal

Since the right triangle is a well-known special case and since the second given mentions right angles, let us establish the proof first for this special triangle (Fig. 3.8). With the third given in mind, let us construct a

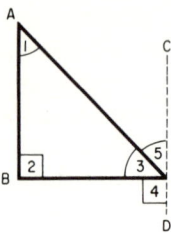

Fig. 3.8 *Right triangle (special case)*

line CD parallel to the vertical side of the triangle. The problem now almost solves itself, as follows:

(1)   Angle 5 = angle 1   Given (3)
(2)   Angle 2 = angle 4   Given (2)
(3)   Angles 4 + 3 + 5 = 180 degrees   Given (1)
(4)   Therefore, angles 2 + 3 + 1 = 180 degrees   Substitution

We have thus solved the special case of right triangles and must now solve the problem for triangles in general. Suppose we take the general triangle shown in Fig. 3.9 and apply the right triangle solution to it. Again we construct a line CD parallel to the opposite side of the triangle. When we apply the logic for the special case to the general triangle (using corresponding interior angles), we find that it solves the general case as well!

If we had attempted to solve the general case first, however, it might not have been obvious to construct that helpful parallel line. In the special case, it is far more likely that we would have considered this construction.

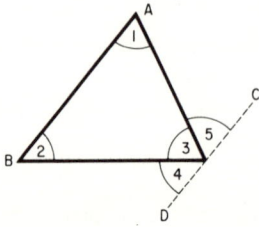

Fig. 3.9 *Any triangle (general case)*

Computer Example

Let us now study a programming problem that uses analogy. With the present emphasis upon energy conservation, you may have been monitoring the fuel consumption in your home on a monthly basis by means of the kilowatt meter and monthly bills. Sample data might be that shown in Table 3.6.

Table 3.6 *Kilowatt Usage*

Date	Meter Reading	Monthly Bill
01/20/79	00231	
02/21/79	02143	$110.30
03/20/79	04498	$120.10
04/19/79	07634	$131.26
05/20/79	11114	$135.71
06/22/79	14705	$140.06
07/19/79	18298	$143.71
08/20/79	21752	$141.62
09/23/79	25153	$135.02
10/19/79	28328	$130.16
11/24/79	31710	$125.14
12/19/79	34618	$119.21

Using this data, you want to determine the following:

1. Kilowatts per dollar on a monthly basis
2. Average kilowatts per dollar
3. Average kilowatts per month
4. Average monthly bill

With a little thought, you realize that you have already solved a very similar problem, the transportation problem in Prescription 1 of Chap. 2. Here you calculated your car's average number of miles per gallon, the inputs being odometer readings and the number of gallons in each refill. The relationship of the two problems in terms of input and output is shown in Table 3.7.

It is clear not only that these problems are analogous but that the same logic can be used to solve both. Their only difference will be the printed messages that are output from the program. We can thus consider these problems to be almost equivalent. The program to solve the kilowatt problem is shown in Example 3.5. With the exception of different variable names and printed messages, it is exactly like that in Example 2.2. The outputs from this program are shown in Table 3.8.

Table 3.7 Analogous Problem Relationships

Transportation Problem	Energy Problem
Input	
(1) Odometer reading	(1) Kilowatt meter reading
(2) Number of gallons filled	(2) Monthly bill
Output	
(1) Average miles per gallon	(1) Average number of kilowatts per dollar
(2) Average miles between refills	(2) Average number of kilowatts per month
(3) Average number of gallons between refills	(3) Average monthly bill

Table 3.8 Output from Energy Program

Kilowatts per Dollar
17.33
19.60
23.89
25.64
25.63
25.00
24.38
25.18
24.39
27.02
24.39

Average kilowatts per dollar: 24.00
Average kilowatts per month: 3126.09
Average monthly bill: 130.20

In summary, analogy is a vital tool in programming, allowing us to perceive situations both of equivalence and similarity. An example of equivalence would be the use of an existing program solution from a programming library.

As already pointed out, programming problems may be similar yet not equivalent in every respect. In such cases, analogy still provides a powerful boost, for the solution of a similar problem may provide insight into the problem at hand, although care must be taken to assure that there is *sufficient* similarity.

Example 3.5 *Energy Program*

```
PROGRAM: TO COMPUTE KILOWATT USAGE

  set TOTAL_KILOWATTS    to   0
  set TOTAL_COST   to   0
  print 'KILOWATTS PER DOLLAR'
  input READING_VALUE, NUM_READINGS

  for CURRENT_READING = 1 to   NUM_READINGS do the following

     input NEXT_READING_VALUE, AMOUNT_BILL
     set KILOWATTS_DOLLAR   to   (NEXT_READING_VALUE - READING_VALUE) /
                                   AMOUNT_BILL
     print KILOWATTS_DOLLAR

     add (NEXT_READING_VALUE -   READING_ VALUE) to TOTAL_KILOWATTS
     add AMOUNT_BILL   to   TOTAL_COST
     set READNG_VALUE   to   NEXT_READING_VALUE

  set AVG_KILOWATJS_DOLLAR   to   TOTAL_KILOWATTS / TOTAL_COST
  print 'AVERAGE KILOWATTS PER DOLLAR:',AVG_KILOWATTS_DOLLAR

  set AVG_KILOWATTS_MONTH   to   TOTAL_KILOWATTS / NUM_READINGS
  print 'AVERAGE KILOWATTS PER MONTH:',AVG_KILOWATTS_MONTH

  set AVG_MONTHLY_BILL   to   TOTAL_COST / NUM_READINGS
  print 'AVERAGE MONTHLY BILL:',AVG_MONTHLY_BILL

end  * program *
```

PRESCRIPTION 8

The Voice of the Majority Is No Proof of Justice

A rite is a ceremonial or formal, solemn act, observance, or procedure in accordance with prescribed rules or customs known as *rituals.*
Often there are strong pressures to conform to these rituals. "Ritualistic
problem solving" is a solution process that conform explicitly and unre-
lentingly to a specified set of rules.

Computers foster the use of procedures, or rituals, because every
phase of a computer system—system logic, user instructions, instructions
for operations personnel, the transmission of data and results—demands
them. Procedures even exist for developing specific computer programs in
terms of program development tools or methods. Procedures also exist for
finding errors in a program and verification of a computer design.

The drawback of ritualistic problem solving is that we can become
so bound by rules and procedures that the solution to a problem is itself
approached as a ritual.

We all practice ritualistic problem solving to some degree in our
daily lives as well as in the programming context. The difficulty is that
rituals can actually become constraints to creative problem solving

whenever problems defy known procedures or guidelines. For example, if imagination cannot be exercised, attempting to abide by rituals can lead to mental blocks. It is therefore imperative that we understand the nature of ritualistic problem solving and resort to creative methods when these are warranted.

Often the pressures to conform are brought about by bandwagon influences. Bandwagon pressures are those induced by a group. They are often exhibited in the behavior of just three individuals. Two of them will tend to share common goals and perspectives, which may then be forced upon the third individual. If we extend this simple case to larger groups, it becomes apparent how bandwagon psychology can influence the problem-solving process.

Group pressures may derive from group rituals or from the preconceived notions of specific individuals within the group. The only solution to this difficulty is to be aware of these bandwagon pressures and make a concerted effort at objectivity. Imagination can be useful here, for creative ideas may halt the bandwagon and provide alternative approaches.

PRESCRIPTION 9

Walk with Wise Men—Communicate

Often a problem or system is sufficiently complicated for it to require more than one programmer to achieve a solution. The usual assumption is that the more men placed on the job, the less time it will take to complete (see Fig. 3.10).

Fig. 3.10 *Perceived number of months to complete a programming project*

However, when a computer problem is very complex and involves a set of tortuous interrelationships, the actual time needed for a solution is better illustrated by Fig. 3.11. Adding more programmers to a project can actually lengthen its schedule, rather than shorten it. It appears that there exists an optimal number of programmers to complete a project in an efficient manner.

Regardless of the programming manpower level, proper communication between individuals is an essential ingredient of a successful program. Communications can affect its completion time, as well as its usability. If groups of programmers are designing various components of a

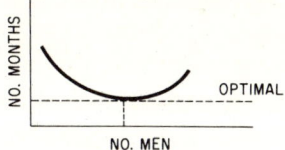

Fig. 3.11 *Actual number of months to complete a programming project*

system in an independent manner, and not communicating properly, the overall solution may be impaired. The disjointed nature of the work process is all too easily reflected in the end product.

The number of communication interrelationships that can be generated is a function of the number of individuals involved. A communication relationship occurs whenever any one individual in a programming group is able to communicate with any other individual. The total number of communication relationships or "links" is given by the binominal coefficient formula:

$$R = \frac{N!}{2!(N-2)!}$$

where:

R = total number of relationships
N = number of programmers involved

Table 3.6 shows how the number of communication relationships can grow exponentially.

Table 3.6 *Total Relationships As a Function of the Number of Individuals*

N	R
2	1
5	10
10	45
25	300
50	1225
75	2775
100	4950

Effective communication in a programming group is therefore imperative if the integrity of the system solution is to be maintained. Improper communication can lead to inadequately defined problem definition, invalid operations, and unachievable solution goals.

PRESCRIPTION 10

All Eggs Can Be Cracked

This prescription says that the solution to a problem often involves finding the "key" that will unlock the necessary operations. The key may be explicitly provided in the givens, or it may be uncovered by logical techniques, such as inference. Without it, a problem may be very difficult or impossible to solve.

Non-Computer Example

Figure 3.12 shows the street layout between a starting point and a destination point. We wish to determine the total number of possible paths from the starting to the destination point. An obvious trial-and-error approach would be to enumerate all paths, but the tedium of drawing and then counting them would soon become apparent.

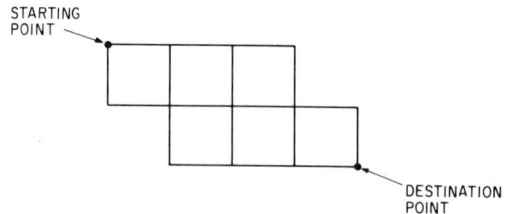

Fig. 3.12   *Street grid problem*

A more practical way of solving the problem, using a "key," is as follows. Put a 1 at the starting point and another 1 at every corner reachable in only one way. Now put a 2 at the first and every other corner reachable in exactly two ways, a 3 at every corner reachable in three ways, and so on. It becomes clear that the number on every corner must be the sum of the one or two nearest numbers along the paths leading to that corner. If we continue in this manner, we eventually solve the original problem. As a matter of fact, we also solve it for any point on the grid relative to the starting point.

Figure 3.13 shows the labeling for the complete street grid and reveals that there are exactly 13 ways to go from the starting point to the destination point. The key insight that provided this numbering system so trivializes the problem that handling even a very large street grid becomes a piece of cake.

A practical application of the grid problem is determining the number of shortest paths that a chess rook can take from one corner of a chessboard to that diagonally opposite. Since a chess rook can move

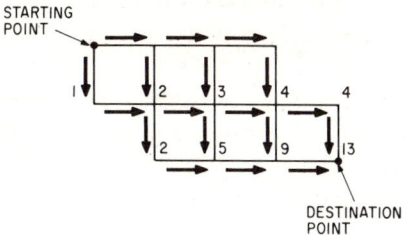

Fig. 3.13 *Solution to street grid problem*

only horizontally or vertically, the shortest paths are those that adjust *each* rook move in a direction, either horizontal or vertical, *toward* the goal.

Since this problem is the same as the street grid problem, we will use the same ''key.'' When the chessboard has been labeled as shown in Fig. 3.14, the labels provide the number of shortest paths from the starting square to any square on the board. The upper right-hand corner of Fig. 3.14 tells us that there are 3,432 ''shortest'' paths.

I	8	36	120	330	792	1716	3432
I	7	28	84	210	462	924	1716
I	6	21	56	126	252	462	792
I	5	15	35	70	126	210	330
I	4	10	20	35	56	84	120
I	3	6	10	15	21	28	36
I	2	3	4	5	6	7	8
R	I	I	I	I	I	I	I

R = INITIAL ROOK POSITION

Fig. 3.14 *Solution to chess problem*

Computer Example

We shall now illustrate the importance of finding ''keys'' to computer problems. The first problem is to find the greatest common divisor of a pair of two numbers, FIRST__NUM and SECOND__NUM. Only the primitive operations of addition, subtraction, multiplication, division, and comparison of integers can be used. At first glance, this constraint would seem to create an impossible situation. However, with some thought, a simple key may be found.

The key is the realization that the greatest common divisor of a pair of positive integers will also be a divisor of their difference unless the latter is zero. This difference will always be smaller than the larger of the two numbers; if it is zero, then the common integer value of the two numbers is also their greatest common divisor. We will thus apply this property and iterate until the difference reaches zero. The method will terminate in a finite number of steps, since the sum of the two numbers is reduced at each iteration, and neither can be reduced to zero. The simple program to solve this problem is shown in Example 3.6.

Example 3.6 *Program to Find Greatest Common Divisor of (M, N)*

```
PROGRAM: TO COMPUTE GREATEST COMMON DENOMINATOR (GCD)

    input FIRST_NUM, SECOND_NUM
    while FIRST_NUM  not equal SECOND_NUM do the following

        if FIRST_NUM greater than SECOND_NUM then

            subtract SECOND_NUM  from FIRST_NUM

        else

            subtract FIRST_NUM  from SECOND_NUM

    set GCD  to  FIRST_NUM
    print 'GREATEST COMMON DIVISOR:',GCD

end  * program *
```

The second problem concerns prime numbers. A prime number is an integer that cannot be divided by any number other than itself or 1 and produce an integer result. For example, 3 is a prime number; 4 is not.

You are asked to write a program to calculate and print all prime numbers between 2 and the integer MAX__INT, a number that is input to the program. Solution of the problem requires a method of determining if a specific number, say N, is prime or not.

Let us assume that all the arithmetic is integer and will provide integer results. For example, if we divide any two nondivisible integers, such as 3 and 2, any remainder is *truncated;* that is, the result is 1. As a matter of fact, this is the key for determining whether any number N is prime.

Our approach is to calculate QUOT, the quotient of N divided by DIVISOR, a trial divisor. If there is no remainder, when we multiply QUOT by DIVISOR, we will obtain N again. On the other hand, if DI-VISOR is not exactly divisible by N, the remainder is lost, and the result will be less than N. For example, if N is 7 and DIVISOR is 3, then:

$$(N/DIVISOR) \times DIVISOR = (7/3) \times 3 = 2 \times 3 = 6$$

Since the result, 6, is less than 7, the two numbers are not divisible. On the other hand, if N is 6 and DIVISOR is 2, then:

(N/DIVISOR) $\times$ DIVISOR = (6/2) $\times$ 2 = 3 $\times$ 2 = 6

Since the result, 6, *does* equal N, the two numbers are divisible.

Now that we know how to determine if two numbers are divisible, we can determine if an integer N between 2 and N is prime by iterating N from 2 to N − 1 using this approach. If at any time during the iteration a number is found that will make the number N divisible, we know that the number N is not prime. On the other hand, if no such number exists, we know that the number N is prime.

The program will iterate from 2 to the input value, MAX_INT, so as to produce all prime numbers from 2 to N − 1; it is shown in Example 3.7.

Example 3.7   *Program to Calculate Prime Numbers*

```
PROGRAM: to COMPUTE PRIME NUMBERS

   input MAX_INT
   set N  to  2
   print 'THE PRIME NUMBERS FROM 2 to', MAX_INT, 'are:'

   while N less or equal MAX_INT do the following

      set DIVISOR  to  2
      while (DIVISOR less or equal N-1 and PROD not N) do the following

         set QUOT  to  integer portion of (N / DIVISOR)
         set PROD  to  integer portion of (QUOT * DIVISOR)
         IF PROD not equal N then

            add 1  to  DIVISOR

      if DIVISOR greater than N-1 then

         print N
      add 1  to  N

end  * program *
```

In summary, problem solutions may "hinge" on very specific circumstances. The problems can often be cracked by finding the key or keys hidden in these circumstances.

PRESCRIPTION 11

Beware of Anxiety—It's Heavy

Another cause of mental blocks is placing too much importance on obtaining a solution as soon as possible. Such an attitude can only create additional stress in the problem-solving process. We often notice this effect in personal problems where the solution is very important to our lives.

It is often difficult to think a problem through if we are uncomfortable for any reason; worrisome distractions disrupt the process of orderly thinking, for the brain has more than one input to deal with. We should therefore concentrate on the problem but attempt to ignore the significance of obtaining its solution. A very powerful approach is temporarily to make a game out of work, even if the problem is very serious.

PRESCRIPTION 12

Umbrellas Are Useful

Defensive problem solving involves the mental attitude that there will always be situations where the chosen solution will fail, situations that the problem never seemed to involve are so sure to crop up.

In a programming context, defensive problem solving may be said to be analogous to "defensive programming." Defensive programming is simply designing a program in a manner that attempts to foresee every possible situation that might arise and that provides the logic to handle all of them, no matter how remote.

Defensive programming mechanisms include the following:

1. Data controls
2. Reasonable tests
3. Checkpoint/restart
4. Protected or privileged modes
5. Diagnostic aids
6. Up-to-date documentation
7. Protection against incorrect input
8. Checking for illegal boundary conditions

An effective way to go about such programming is for the programmer to assume that the program design and code will be extensively reviewed by others. Therefore, at every point during problem definition, design, and coding, the programmer tries to think of everything that might make the design review easier. A program should be designed and coded as if the programmer were responsible for it *forever*.

Computer Example

The importance of defensive programming can be demonstrated as follows. Research biologists at the Biological Research Institute make wide use of a mathematical relationship known as the quadratic equation, which is expressed as follows:

$$AX^2 + BX + C = 0$$

where A, B, and C are constants known by the biologists. The variable the biologists are interested in, X, has two solutions, the roots of the quadratic equation, which are expressed as follows:

$$X_1 = \frac{-B + \sqrt{B^2 - 4AC}}{2A} \qquad X_2 = \frac{-B - \sqrt{B^2 - 4AC}}{2A}$$

Imagine that you are a newly hired programmer for the Institute and are asked to write a program to solve quadratic equations with data provided by the biologists. Input consists of a deck of cards, each record containing the variables A, B, and C. Output consists of the solution roots associated with each set of variables. Since the program appears trivial, the program solution, shown in Example 3.8, is coded quickly. After some initial testing, you turn over the program to the biologists.

Example 3.8 *Biology Program (First Attempt)*

```
PROGRAM: TO COMPUTE SOLUTION ROOTS

Constant Definitions

  NUM_EQUATIONS = 50

  print 'COEFICIENTS:      ROOTS:'
  for CURRENT_EQUATION = 1 to NUM_EQUATIONS do the following

    input A, B, C
                                       2
    set ROOT1  to  (-B + square root of (B - 4*A*C)) / (2*A)
                                       2
    set ROOT2  to  (-B - square root of (B - 4*A*C)) / (2*A)
    print 'A:',A,'B:',B,'C:',C,'ROOT1:',ROOT1,'ROOT2:',ROOT2

end  * program *
```

On two separate occasions the program does not work properly and prints mysterious results. Table 3.7 shows the input data used when the program failed.

Table 3.7 *Input Data*

Case	Coefficient Input Data		
	(A)	*(B)*	*(C)*
1	0	6	2
2	9	2	3

At your desk, your review of the program logic cannot find anything wrong with the way you coded the quadratic formula. Nevertheless, when you plug the constant data for each error case into the quadratic formula, Table 3.8 results.

Table 3.8 *Computed Results*

Case	Computed results	
	Root 1	*Root 2*
1	(−6 + SQUARE ROOT OF (36))/0	(−6 − SQUARE ROOT OF (36))/0
2	(−2 + SQUARE ROOT OF (−104))/18	(−2 − SQUARE ROOT OF (−104))/18

From your mathematics, you realize what caused the problem. In the first case, you attempted to divide the numerator by zero, an operation that is undefined. In the second case, you attempted to find the square root of a negative number, which has no *real* solution roots.

You had not anticipated these two situations and had not practiced defensive programming. In particular, you violated mechanism (7) of defensive programming. It is therefore necessary to rewrite your program to take these invalid input situations into consideration, as illustrated by Example 3.9.

Example 3.9 *Biology Program (Final Version)*

```
PROGRAM: TO COMPUTE SOLUTION ROOTS

Constant Definitions
  NUM_EQUATIONS = 50

  print 'COEFICIENTS:     ROOTS:'
  for CURRENT_EQUATION = 1 to NUM_EQUATIONS do the following

    input A, B, C
                      2
    if A not equal 0 and (B - 4*A*C) greater or equal 0 then

                             2
      set ROOT1  to  (-B + square root of (B - 4*A*C)) / (2*A)
                             2
      set ROOT2  to  (-B - square root of (B - 4*A*C)) / (2*A)
      print 'A:',A,'B:',B,'C:',C,'ROOT1:',ROOT1,'ROOT2:',ROOT2

    else

      print 'ILL-FORMED EQUATION A:',A,'B:',B,'C:',C

end  * program *
```

The defensive logic added to the program checks whether variable A is nonzero and the expression $(B^2 - 4AC)$ is greater than zero.

In summary, program errors are often caused by a lack of defensive programming. Defensive programming is essential to a well-written program to protect it from the consequences of unanticipated problems.

PRESCRIPTION 13

He Who Can at All Times Sacrifice Pleasure to Duty Approaches Sublimity

Our mental attitudes determine how easily we are able to solve a given problem. For example, trying to solve a puzzle, we are not likely to have much difficulty so long as our attitude remains pleasurable. In fact, we may even enjoy solving the puzzle so much that we are disappointed when it is solved!

Hence, an important factor influencing problem solving is the role that games play. It should be noted that the "game effect" is double-edged. On the one hand, it can spark interest in the problem and happily spur us on to a solution; on the other, it may be detrimental. The enjoyment of the game may start to interfere with the technical handling of the problem. It is therefore imperative that we be aware of the game effect and guard against its perils.

4

Solving Larger Problems

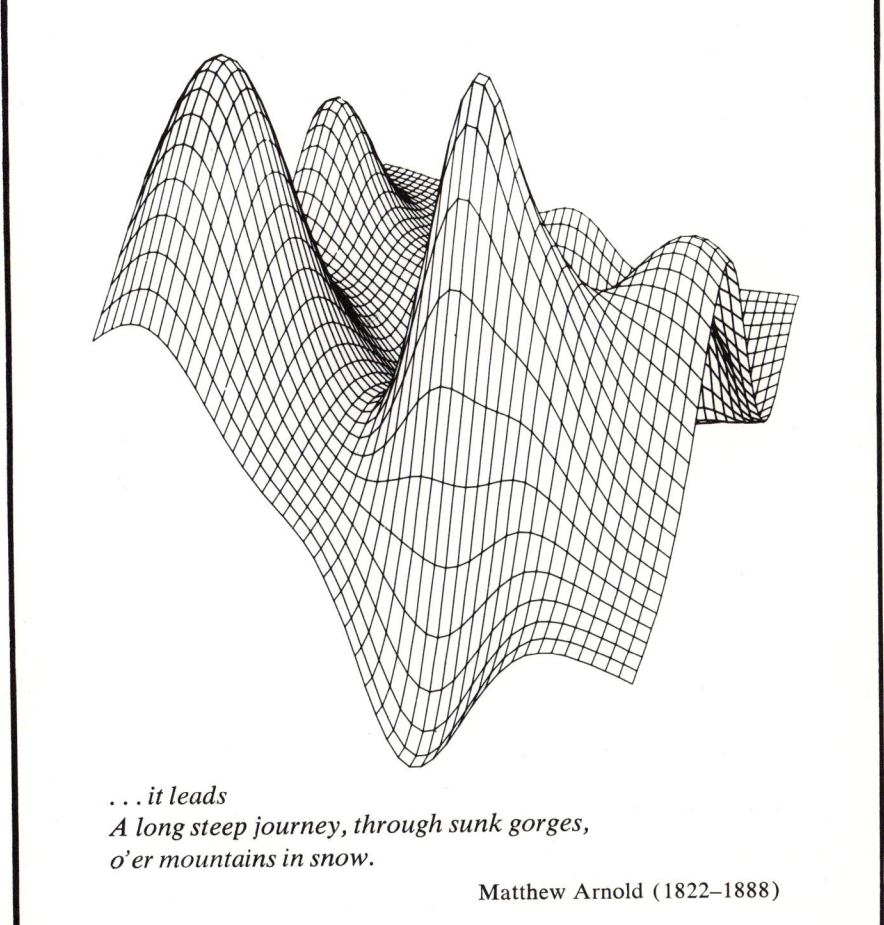

. . . it leads
A long steep journey, through sunk gorges,
o'er mountains in snow.

Matthew Arnold (1822–1888)

4.1   Introduction

Before writing a computer program to solve a problem, you must develop an algorithm. Algorithms are a kind of recipe that spells out explicitly the steps to be carried out by the computer. They are not always easy to generate, especially if many steps are involved. In fact, when introduced to a problem, you may be swamped by the mounds of details to be considered. If you attempt to digest them all at once, you are likely to achieve nothing but cramps.

An approach for dealing with this kind of complexity says that you should always start at the *top* and deal with the general characteristics of the problem first. Your ever increasing understanding of the problem will then enable you to consider the details more intelligently.

4.2   Tree Structures

A convenient representation of this progression from the general to the specific is the tree structure shown in Fig. 4.1.

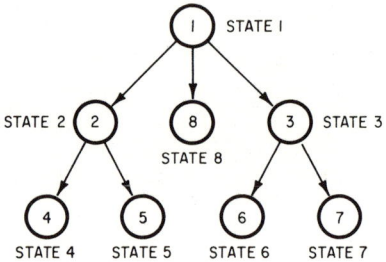

Fig. 4.1   *General tree structure*

The tree, whose name originates from the real thing, is a very useful tool for representing logic. Compared to the real-world tree, it has an inverted structure, with the root at the top and the culminating branches at the bottom. The circles are called "nodes," or "states," and the lines are called "branches." The terminating nodes, which do not connect to the other nodes, are called "leaf nodes."

You will notice that the tree structure in Fig. 4.1 is comprised of eight states. Starting at state 1, there are three branches—to states 2, 8, and 3, respectively. Branches represent the possibility of different actions, these actions leading from one state to another. For example, to arrive at state 5, we must begin at state 1, proceed to state 2, and from there jump to state 5. Each state represents a unique set of conditions. To put it another way, a given state can be obtained only if a specific set of operations occur. A state in a programming problem represents our current level of understanding of the problem; the further down the tree we go, the more detailed the information.

A simple way to illustrate the use of a tree structure is to consider the beam balance in Fig. 4.2. If two objects, A and B, are placed on a beam support at its middle, three things can happen. If A is greater than B, the balance will tip, the left side will go down, and the right side, up. If B is greater than A, the right side will go down and the left side, up. If A and B are equal, both sides will remain as they are. The three possible states are shown in Fig. 4.3.

Fig. 4.2   *Beam balance*

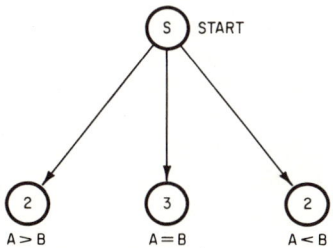

Fig. 4.3   *States of a tree*

Non-Computer Example

The following problem will illustrate the use of trees to represent logic as well as the top-down approach to problem solving.

Given five pennies, one of which is an odd weight, either lighter or heavier than the other four pennies, determine, in three weighings, which penny is the odd penny and whether it is heavier or lighter. The most general problem description is: Determine the odd penny. A general plan to solve this problem is shown in Fig. 4.4.

Conceptually, the first weighing of two coins against two others provides us with a great deal of information, depending upon whether the scale tips to the left, balances, or tips to the right.

If the scale balances, we have narrowed the search to one odd penny; on the second weighing, this penny can be weighed against a good penny to determine if it is lighter or heavier.

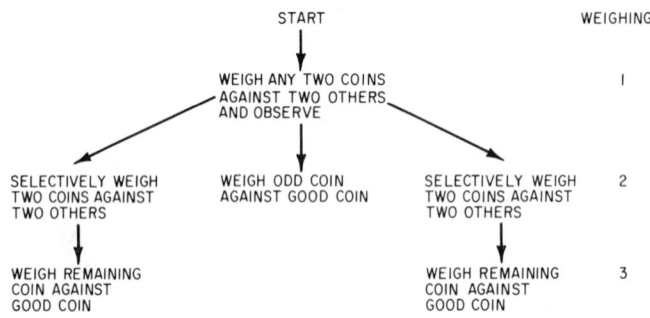

Fig. 4.4   *General plan to solve coin problem*

If the scale does not balance, it has tipped either to the left or right. If we somehow switch the coins and again weigh two coins against two others, we gain still more information. In fact, after this weighing, only two possibilities remain: the odd penny is either lighter or heavier. The third weighing against a known good penny will decide.

We now apply the top-down approach by developing a detailed refinement of the general plan described above. Let us use circled numbers to indicate the specific pennies to be weighed and arrows to indicate whether the scale tipped to the left or to the right or balanced. The symbol indicating whether the penny is lighter or heavier, respectively, is the penny number subscripted by the letter L or H.

We start by first weighing any four pennies, two on each side of the beam balance, as shown in Fig. 4.5. If the two sides balance, indicated by an equals sign, then penny 5 must be either heavier or lighter. However, if the left side goes down, indicated by the left arrow, then we know that either 1 or 2 is heavier or 3 or 4 is lighter. Likewise, if the right side goes down, we know that either 3 or 4 is heavier or that 1 or 2 is lighter.

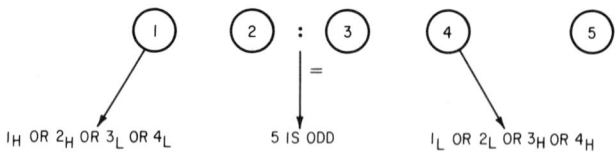

Fig. 4.5   *Results of first weighing*

Let us suppose that the scale balances in the first weighing. How can we determine if penny 5 is heavier or lighter? Simply by weighing it against any normal penny K—in this case, any of the other four pennies now known to be standard—as shown in Fig. 4.6. Since penny 5 is odd, it

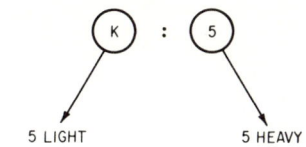

Fig. 4.6 *Results of second weighing*

will not balance with good penny K. If the right side goes down, then 5 is heavier; if the left side, lighter.

Now let us consider the two possible results of the first weighing in which a balance does *not* occur (a tilt to the left or right). The imbalance assures us that penny 5 is good. You will note that either result reduces the number of possible cases to four. Before pennies 1, 2, 3, and 4 were weighed, there were eight possible cases involving the odd penny: Any of the four could be heavier, or any of the four could be lighter. That is, the probability is now cut in half.

Let us first consider the case where the left side goes down, indicated by the left arrow. We then know that either penny 1 or 2 is heavier or penny 3 or 4 is lighter. How can we go about determining the odd penny?

Essentially, there are three things we can do with any given penny. We can leave it where it is, switch it to the other side of the scale, or remove it. This being the case, and there being four pennies, we have many possible ways in which to proceed. Let us consider one of them.

Remember that this will be the second weighing and only one more is permitted. Let us remove penny 2 and switch penny 3 to the left side. We then take penny 5, known to be of normal weight, and place it on the right side. The new situation is illustrated in Fig. 4.7.

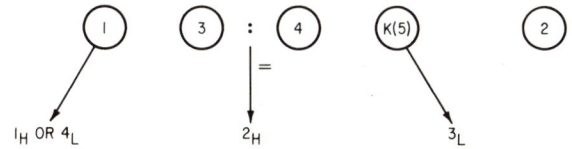

Fig. 4.7 *Results of second weighing*

Again only three things can happen. (1) If the beam balances, the odd penny has been removed. The odd penny is therefore penny 2, and the previous weighing proved that it was heavier, not lighter. If the pennies do not balance, then penny 2 is normal. (2) If the scale tips to the left, then only the pennies not switched could influence the balance. Since moving penny 3 didn't change anything, it is normal, and we know that either penny 1 is heavier or penny 4 is lighter. (3) If the balance tips to the right, it is due to our switches. Since only penny 3 was switched, it has to be the odd penny, and we already know that it is lighter.

We have now only to determine whether penny 1 or 4 is lighter or heavier. To do so, all we need to do is weigh penny 1 against a normal penny to determine if it is odd, as shown in Fig. 4.8. If the beam tips to the left, we know that penny 1 is heavier; if it balances, we know that penny 4 is lighter.

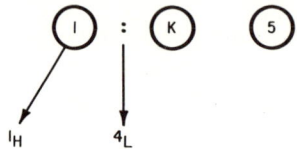

Fig. 4.8 *Results of third weighing*

We have not solved the entire problem. We must now consider the case when the right side goes down in the first weighing and provide a further refinement. Remember that in this case either penny 1 or 2 is lighter or penny 3 or 4 is heavier.

First of all, switch pennies 2 and 3, remove penny 4, and put penny 5 in its place, as shown in Fig. 4.9. If the scale balances, then penny 4 is heavier. If the balance remains as it was—tilted to the right—we know

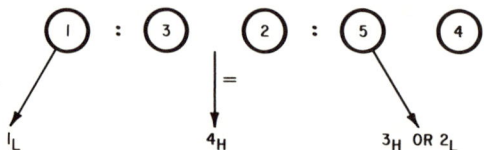

Fig. 4.9 *Results of second weighing*

that only penny 1 remained in position and therefore has to be lighter. However, if the balance changes, then either penny 3 is heavier or penny 2 is lighter. To determine which, let us weigh penny 2 against a normal penny, as shown in Fig. 4.10. If the beam tips to the right, we know that penny 2 is lighter; if it balances, we know that penny 3 is heavier.

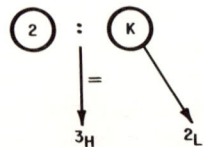

Fig. 4.10 *Results of third weighing*

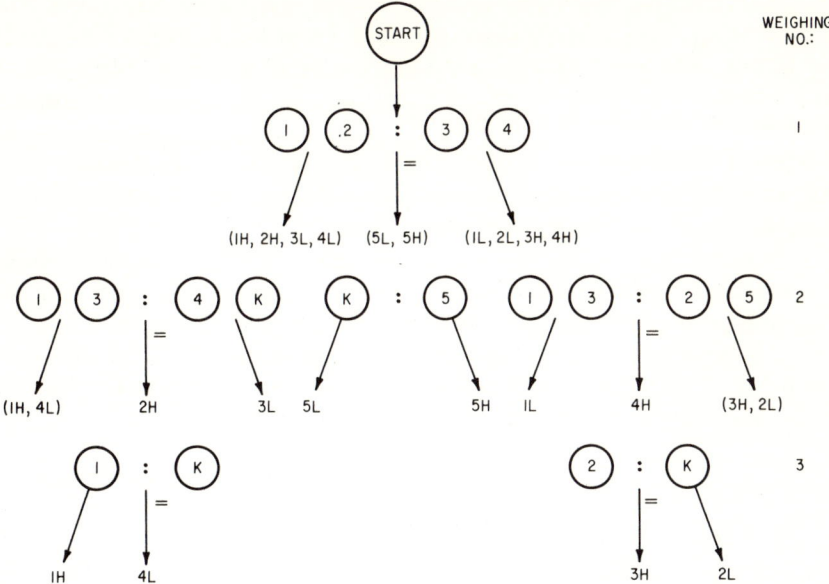

Fig. 4.11   *Final tree solution to coin problem*

This step completes the analysis of the problem. If we now inter-connect all the details, we obtain the tree structure shown in Fig. 4.11.

In Fig. 4.4, we defined a general plan for solving the coin-balancing problem. After each weighing, we added additional detail to the original plan until we obtained a detailed algorithm for solving the problem.

4.3   Top-Down Programming Process

The top-down approach to problem solving is similar to the method-ology used for the coin-balancing problem and can be represented graphi-cally by tree structures. Figure 4.12 illustrates the overall structure.

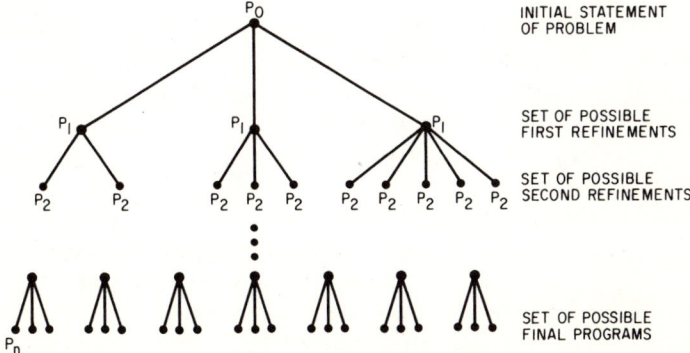

Fig. 4.12   *Overall structure of top-down programming process*

The top-down approach structures a problem as a set of levels. The top-most level, P_0, comprises the most general statement about the problem. Below this level, the branches of the tree represent various programming design decisions. Each successive level represents a further refinement of the problem toward a final solution. At each level, the programmer must choose among the alternative branches, or solutions, leading to the final detailed solution. As more and more information is obtained, the overall design becomes clearer.

Figure 4.13 illustrates the tree structure of a programming problem requiring four levels of refinement. The most general statement of the problem is at the top, and the detailed solution at the bottom.

The nodes represent the program components at each level of refinement. The linkage between levels represents the break-up of the design into subsections for further refinement.

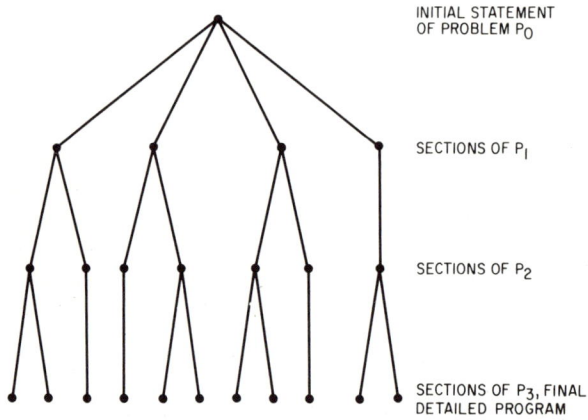

INITIAL STATEMENT
OF PROBLEM P_0

SECTIONS OF P_1

SECTIONS OF P_2

SECTIONS OF P_3, FINAL
DETAILED PROGRAM

Fig. 4.13 *Top-down structure of a specific program*

Non-Computer Example

As an illustration of the top-down approach, consider the problem of fixing a martini. A graphical representation is shown in Fig. 4.14.

The highest level description of the problem, P_0, is: Fix a martini. The next step is to break the problem down with further refinement, resulting in level P_1, which consists of freezing ice, mixing ingredients, and evaluating the results.

Let us look at the first task of freezing ice. In fact, let's consider the next level of refinement, P_2, that is associated with this task. We must obtain an ice tray, fill it with water, and place it into the freezer. Another component of level P_1—mixing ingredients—is refined in a similar manner until level P_2 is achieved. The final component of level P_1—evaluating the results—is also refined to level P_2.

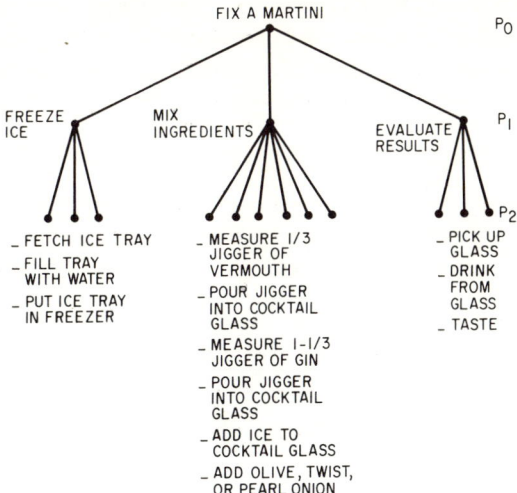

Fig. 4.14 *Top-down design to fix a martini*

This simple example illustrates the general approach to top-down programming. While rather artificial, it does provide a "flavor" to the concept.

The highest and most general description of the problem is stated first. Next, the problem is broken down into more detail in the next level of refinement, and this process continues until a final detailed description of the solution is achieved.

4.4 Computer Examples

The top-down approach is a widely applied technique in programming that provides a useful guide for the programmer. Since it is independent of any particular programming language, a program may be designed before the language or even the computer hardware is chosen.

We will now examine two programming problems. The first, which is very simple, is programmed in five languages. The second is more difficult and involves inventory control.

Sales Commission

You are asked to write a program to calculate sales commissions for a group of salesmen employed by the Topdown Vacuum Cleaner Company. Commissions are to be computed for sales in excess of quotas; if quotas are not met, no bonus is paid. In addition, the program must determine the largest commission amount and the number of the associated salesman.

To define the problem clearly, our first step is to define the specific input and output formats, as shown in Table 4.1. For each salesman input record, the program is to output a printed line as defined by the output format. The sales commission amount is to be computed as follows:

Commission = Bonus factor × (Actual sales − Sales quota)

Table 4.1 *Sales Commission Input/Output Formats*

Inputs A sequence of salesman records each with the following data:

Item	Card Column Positions
Salesman number (dddddddd)	1–8
Quota Amount (ddd.dd)	10–15
Actual Sales Amount (ddd.dd)	17–22
Bonus Factor (.dd)	24–26
Where:	
d = a digit	

Outputs A Sales Commission Report as follows:

Col 22
↓

SALES COMMISSION REPORT

Col 6 ↓ Salesman Number	Col 19 ↓ Sales Quota	Col 29 ↓ Actual Sales	Col 40 ↓ Bonus Factor	Col 51 ↓ Commission Paid
59291259	400.10	380.05	.20	0.00
92047193	300.00	340.55	.18	7.30
.	.	.	.	.
.	.	.	.	.

Largest Sales Commission: 7.30 sold by salesman No. 92047193

When we apply the top-down approach, the problem statement at the topmost general level is defined as follows:

$$P_0$$
Calculate Sales Commissions

It is now necessary to consider the overall structure of the problem and refine the problem statement with a little detail—the basic order of the calculations and the general condition under which the program terminates. Level P_1 results, as shown in Example 4.1.

Example 4.1   *P₁ Level of Sales Commission Report*

```
PROGRAM: TO COMPUTE SALES REPORT
    while more data do the following
        input record
        process data
        print results
    print largest commission paid
end * program *
```

Note that P_1 is a more refined statement of the problem that is basically described in English. Now that we have the basic program logic structure, we are in a position to refine level P_1 with more detail, as shown in Example 4.2.

Example 4.2   *P₂ Level of Sales Commission Report*

```
PROGRAM: TO COMPUTE SALES REPORT
    print report headings
    set largest commission to 0
    while more data do the following
        if commission paid this salesman then
            compute commission
            determine largest commission
        else
            set no commission paid this salesman
    print largest commission paid
end * program *
```

We will now refine the problem with one more level of detail that provides the specific inputs, outputs, and the specific condition under which the program terminates, as in shown Example 4.3.

There is no need to refine the problem beyond level P_3, for this level is very detailed. The program can now be coded in any chosen programming language.

In summary, we started out with the highest general description of

Example 4.3 P_3 Level of Sales Commission Report

```
PROGRAM: TO COMPUTE SALES COMMISSIONS

    print '                    SALES COMMISSION REPORT'
    print 5 BLANK LINES
    print '      SALESMAN     SALES      ACTUAL     BONUS      COMMISSION'
    print '      NUMBER       QUOTA      SALES      FACTOR     PAID'
    print '      .......      .....      ......     ......     ..........'
    print 2 BLANK LINES

    set MAXCOMMISSION  to  0
    input SALESMAN
    while SALESMAN not zero do the following

       input QUOTA, SALES, BONUSFACTOR
       if SALES greater than QUOTA THEN

          set COMMISSION  to  BONUSFACTOR * (SALES - QUOTA)

          if COMMISSION greater than MAXCOMMISSION then

             set TOPSALESMAN  to  SALESMAN
             set MAXCOMMISSIOM  to  COMMISSION

       else

          set COMMISSION  to  0

       print SALESMAN, QUOTA, SALES, BONUSFACTOR, COMMISSION
       print 1 BLANK LINE
       input SALESMAN

    print 'LARGEST SALES COMMISSION: ';MAXCOMMISSION;
          ' SOLD BY SALESMAN NUMBER ';TOPSALESMAN

end  * program *
```

the problem. Subsequent levels were refined with more and more detail until we finally reached level P_3.

Three characteristics of top-down programming are important enough to warrant a brief discussion. First of all, the problem must be clearly understood and defined before actual programming is begun. It is silly to start programming before you have completely grasped the problem.

Second, top-down design is both machine- and language-independent. You must always be more concerned with the overall program logic and design than with the characteristics of any given programming language.

Third, you should debug the program at each level so that the refinements will be correct in relation to previously designed levels. If each level is thoroughly examined for its logical correctness, the need for backtracking to higher levels will be minimized.

The sales commission problem in level P_3 is programmed in the following languages—BASIC, Pascal, PL/1, FORTRAN, and COBOL—in Examples 4.4, 4.5, 4.6, 4.7, and 4.8, respectively.

Example 4.4 *Sales Commission Problem Solved in BASIC (IBM 5110)*

```
0100 REM   **
0110 REM   **    PROGRAM TITLE: SALES COMMISSION REPORT
0120 REM   **
0130 REM   **    AUTHOR: WILLIAM E. LEWIS
0140 REM   **
0150 REM   **    PROGRAM SUMMARY:
0160 REM   **
'0170 REM   **       THIS PROGRAM PROCESSES A SERIES OF SALESMAN RECORDS
0180 REM   **       AND PRODUCES A SALES REPORT BY SALESMAN
0190 REM   **
0200 REM   **       THE LARGEST COMMISSION AMOUNT PAID AND ASSOCIATED
0210 REM   **       SALESMAN NUMBER ARE ALSO COMPUTED AND OUTPUT
0220 REM   **
0230 REM   **       THE PROGRAM TERMINATES WHEN A SALESMAN NUMBER OF ZERO
0240 REM   **       IS DETECTED
0250 REM   **
0260 REM   **
0270            GOSUB 0470
0280            LET M = 0
0290            READ S
0300            IF S = 0 GOTO 0430
0310               READ Q, A, B
0320               IF A <= Q GOTO 0390
0330                  LET C = B * (A - Q)
0340                  IF C <= M GOTO 0380
0350                     LET T = S
0360                     LET M = C
0370                     GOTO 0400
0380 REM
0390                     LET C = 0
0400               PRINT S, Q, A, B, C
0410               READ S
0420               GOTO 0300
0430            PRINT
0440            PRINT 'LARGEST SALES COMMISSION: ';M;' SOLD BY';
0450            PRINT ' SALESMAN NUMBER ';T
0460            STOP
0470 REM
0480            PRINT '                           SALES COMMISSION REPORT'
0490            PRINT
0500            PRINT
0510            PRINT
0520            PRINT
0530            PRINT
0540            PRINT 'SALESMAN','SALES','ACTUAL','BONUS','COMMISSION'
0550            PRINT 'NUMBER','QUOTA','SALES','FACTOR','PAID'
0560            PRINT '........','.....','......','......','..........'
0570            PRINT
0580            PRINT
0590            RETURN
0600            DATA 59291259, 400.10, 380.05, .20
0610            DATA 92047193, 300.00, 340.55, .18
0620            DATA 0
0630            END
```

Example 4.5 *Sales Commission Problem Solved in Pascal*

```
(*=======================================================================*)
(*                                                                       *)
(* PROGRAM TITLE: SALES COMMISSION REPORT                                *)
(*                                                                       *)
(* AUTHOR: WILLIAM E. LEWIS                                              *)
(*                                                                       *)
(* PROGRAM SUMMARY:                                                      *)
(*                                                                       *)
(*     THIS PROGRAM PROCESSES A SERIES OF SALESMEN RECORDS               *)
```

```
(*    AND PRODUCES A SALES REPORT BY SALESMAN                          *)
(*                                                                      *)
(*    THE LARGEST COMMISSION AMOUNT PAID AND ASSOCIATED                 *)
(*    SALESMAN NUMBER ARE ALSO COMPUTED AND OUTPUT                      *)
(*                                                                      *)
(*    THE PROGRAM TERMINATES WHEN A SALESMAN NUMBER OF ZERO IS          *)
(*    DETECTED.                                                         *)
(*                                                                      *)
(*                                                                      *)
(*====================================================================*)

PROGRAM SALES (INPUT, OUTPUT);
VAR
  SALESMAN, TOPSALESMAN    : INTEGER;
  QUOTA,  SALES,  BONUSFACTOR,
  COMMISSION, MAXCOMMISSION : REAL;

PROCEDURE PRINTHEADING;
  BEGIN
    WRITELN(' ':21,'SALES COMMISSION REPORT');
    WRITELN;
    WRITELN;
    WRITELN;
    WRITELN;
    WRITELN;
    WRITELN('     SALESMAN      SALES      ACTUAL      BONUS',
            '   COMMISSION');
    WRITELN('     NUMBER        QUOTA      SALES       FACTOR',
            '   PAID');
    WRITELN('     ........      .....      ......      ......',
            '   ..........');
    WRITELN;
    WRITELN;
  END;


BEGIN
  PRINTHEADING;
  MAXCOMMISSION := 0;
  READ(SALESMAN);
  WHILE SALESMAN <> 0 DO
    BEGIN
      READ(QUOTA, SALES, BONUSFACTOR);
      IF SALES > QUOTA THEN
        BEGIN
          COMMISSION := BONUSFACTOR * (SALES - QUOTA);
          IF COMMISSION > MAXCOMMISSION THEN
            BEGIN
              TOPSALESMAN   := SALESMAN;
              MAXCOMMISSION := COMMISSION
            END
        END

      ELSE

        COMMISSION := 0;

      WRITELN(SALESMAN:13, QUOTA:11:2, SALES:10:2,
              BONUSFACTOR:8:2, COMMISSION:14:2);
      WRITELN;
      READ(SALESMAN)
    END; (*WHILE*)

  WRITELN;
  WRITELN(' ':5,'LARGEST SALES COMMISSION: ';MAXCOMMISSION;
          ' SOLD BY SALESMAN NUMBER ';TOPSALESMAN);

END.
```

Example 4.6   *Sales Commission Problem Solved in PL/1 (IBM 370)*

```
SALES:
  PROCEDURE OPTIONS(MAIN);
  /******************************************************************/
  /*                                                              */
  /*    PROGRAM TITLE: SALES COMMISSION REPORT                    */
  /*                                                              */
  /*    AUTHOR: WILLIAM E. LEWIS                                  */
  /*                                                              */
  /*    PROGRAM SUMMARY:                                          */
  /*                                                              */
  /*       THIS PROGRAM PROCESSES A SERIES OF SALESMAN RECORDS    */
  /*       AND PROCESSES A SALES REPORT BY SALESMAN               */
  /*                                                              */
  /*       THE LARGEST COMMISSION AMOUNT AND ASSOCIATED           */
  /*       SALESMAN NUMBER ARE ALSO COMPUTED AND OUTPUT           */
  /*                                                              */
  /*       THE PROGRAM TERMINATES WHEN A SALESMAN NUMBER OF ZERO IS */
  /*       DETECTED.                                              */
  /*                                                              */
  /*                                                              */
  /******************************************************************/

  DCL (QUOTA,SALES,BIGCOM,COMMIS) FIXED DECIMAL (6,2);
  DCL FACTOR FIXED DECIMAL (3,2);
  DCL (MANNO,MANBIG) FIXED DECIMAL (8);

  PUT EDIT ('SALES COMMISSION REPORT') (COLUMN(22),A(23));
  PUT SKIP(5);

  PUT EDIT ('SALESMAN','SALES','ACTUAL','BONUS',
     'COMMISSION') (COLUMN(6),A(8),COLUMN(19),A(5),COLUMN(29),
     A(6),COLUMN(40),A(5),COLUMN(51),A(10));

  PUT SKIP;
  PUT EDIT ('NUMBER','QUOTA','SALES','FACTOR',
     'PAID') (COLUMN(6),A(6),COLUMN(19),A(5),COLUMN(29),
     A(5),COLUMN(40),A(6),COLUMN(51),A(4));

  PUT EDIT ('.........','.....','......','......',
     '..........') (COLUMN(6),A(8),COLUMN(19),A(5),COLUMN(29),
     A(6),COLUMN(40),A(6),COLUMN(51),A(10));

  PUT SKIP(2);
  BIGCOM = 0.0;
  GET EDIT (MANNO) (F(8));

  DO WHILE (MANNO > 0);

      GET SKIP;
      GET EDIT (QUOTA,SALES,FACTOR) (F(6,2),X(1),F(6,2),
          X(1),F(3,2));
      GET SKIP;

      IF (SALES-QUOTA) >0 THEN DO;
          COMMIS = FACTOR * (SALES - QUOTA);

          IF COMMIS > BIGCOM THEN DO;
              MANBIG = MANNO;
              BIGCOM = COMMIS;

      END;
  END;
  ELSE
      COMMIS = 0.0;
```

```
       PUT EDIT (MANNO,QUOTA,SALES,FACTOR,
           COMMIS) (COLUMN(6),F(8),COLUMN(19),F(6,2),COLUMN(29),
           F(6,2),COLUMN(40),F(3,2),COLUMN(51),F(6,2));
       PUT SKIP;

       GET EDIT (MANNO) (F(8));
   END;

   PUT SKIP(1);
   PUT EDIT ('LARGEST SALES COMMISSION: ') (COLUMN(6),A(26)),
       (BIGCOM) (COLUMN(32),F(6,2) (' SOLD BY SALESMAN NUMBER ')
       (COLUMN(38),A(25)), (MANBIG) (COLUMN(63),F(8)));
END;
```

Example 4.7   *Sales Commission Problem Solved in FORTRAN (IBM 370)*

```
*   **
*   **
*   **    PROGRAM TITLE: SALES COMMISSION REPORT
*   **
*   **    AUTHOR: WILLIAM E. LEWIS
*   **
*   **    PROGRAM SUMMARY:
*   **
*   **        THIS PROGRAM PROCESSES A SERIES OF SALESMAN RECORDS
*   **        AND PRODUCES A SALES REPORT BY SALESMAN
*   **
*   **        THE LARGEST COMMISSION AMOUNT PAID AND ASSOCIATED
*   **        SALESMAN NUMBER ARE ALSO COMPUTED AND OUTPUT
*   **
*   **        THE PROGRAM TERMINATES WHEN A SALESMAN NUMBER OF ZERO
*   **        IS DETECTED
*   **
*   **

       INTEGER SLSMAN, TOPSAL
       REAL QUOTA, SALES, BONUSF
       REAL COMMIS, MAXCOM

       CALL HEADER
       MAXCOM = 0
       READ(1,100) SLSMAN
    10 IF (SLSMAN .NE. 0)
     +    THEN

            READ(1,150) QUOTA, SALES, BONUSF
            IF (SALES .GT. QUOTA)
     +          THEN

                COMMIS = BONUSF * (SALES - QUOTA)
                IF (COMMIS .GT. MAXCOM)
     +              THEN

                    TOPSAL = SLSMAN
                    MAXCOMM = COMMIS

                END IF

            ELSE

                COMMIS = 0
```

```
      END IF

      WRITE(2,200) SLSMAN, QUOTA, SALES, BONUSF, COMMIS
      READ(1,100) SLSMAN
      GOTO 10

      END IF
      WRITE(2,250) MAXCOM

100   FORMAT(I8)
150   FORMAT(2F7.2, F4.2)
200   FORMAT(I13, F11.2, F10.2, F8.2, F14.2)
250   FORMAT('     LARGEST SALES COMMISSION: ',
     +       F5.2, ' SOLD BY SALESMAN NUMBER: ', TOPSAL)

      END

      SUBROUTINE HEADER
      WRITE(2,450)
      WRITE(2,500)
      RETURN
450   FORMAT('                    SALES COMMISSION REPORT' /////)
500   FORMAT('SALESMAN     SALES     ACTUAL     BONUS     COMMISSION' /
     +       'NUMBER       QUOTA     SALES      FACTOR    PAID' /
     +       '........     .....     ......     ......    ..........' //)
      END
```

Example 4.8 *Sales Commission Problem Solved in COBOL (CDC Cyber 74)*

```
      IDENTIFICATION DIVISION.
      PROGRAM-ID.      SALESMAN-REPORT.

      REMARKS. PROGRAM TITLE: SALES COMMISSION REPORT

              AUTHOR: WILLIAM E. LEWIS

              PROGRAM SUMMARY:

                  THIS PROGRAM PROCESSES A SERIES OF SALESMAN RECORDS
                  AND PRODUCES A SALES REPORT BY SALESMAN

                  THE LARGEST COMMISSION AMOUNT PAID AND ASSOCIATED
                  SALESMAN NUMBER ARE COMPUTED AND OUTPUT

                  THE PROGRAM TERMINATES WHEN A SALESMAN NUMBER OF
                  ZERO IS DETECTED


      END OF REMARKS.

      ENVIRONMENT DIVISION.
      CONFIGURATION SECTION.
      SOURCE-COMPUTER.
      OBJECT-COMPUTER.
      SPECIAL-NAMES.
          "1" IS TO-TOP-OF-PAGE.

      INPUT-OUTPUT SECTION.
      FILE-CONTROL.
          SELECT INPUT-SALES-FILE  ASSIGN TO "INPUT"
          SELECT REPORT-FILE ASSIGN TO "OUTPUT"
```

```
DATA DIVISION.
FILE SECTION.
FD   INPUT-SALES-FILE
     LABEL RECORDS ARE OMITTED.

01   SALESMAN-NO.                        PIC 9(8).

01   SALES-INFO.
     05   SALES-QUOTA                    PIC 999V99.
     05   FILLER                         PIC X.
     05   SALES                          PIC 999V99
     05   FILLER                         PIC X.
     05   BONUS-FACTOR                   PIC V99.

FD   REPORT-FILE
     LABEL RECORDS ARE OMITTED.
01   REPORT-RECORD                       PIC X(132).

WORKING-STORAGE SECTION.
01   HEADING-LINE-1.
     05   FILLER                         PIC X(21)    VALUE SPACES.
     05   FILLER                         PIC X(23)
          VALUE "SALES COMMISSION REPORT".

01   HEADING-LINE-2.
     05 FILLER                           PIC X(5)     VALUE SPACES.
     05 FILLER                           PIC X(13)
          VALUE "SALESMAN".
     05 FILLER                           PIC X(10)
          VALUE "SALES".
     05 FILLER                           PIC X(11)
          VALUE "ACTUAL".
     05 FILLER                           PIC X(11)
          VALUE "BONUS".
     05 FILLER                           PIC X(10)
          VALUE "COMMISSION".

01   HEADING-LINE-3.
     05 FILLER                           PIC X(5)     VALUE SPACES.
     05 FILLER                           PIC X(13)
          VALUE "NUMBER".
     05 FILLER                           PIC X(10)
          VALUE "QUOTA".
     05 FILLER                           PIC X(11)
          VALUE "SALES".
     05 FILLER                           PIC X(11)
          VALUE "FACTOR".
     05 FILLER                           PIC X(10)
          VALUE "PAID".

01   HEADING-LINE-4.
     05   FILLER                         PIC X(5)     VALUE SPACES.
     05   FILLER                         PIC X(13)
          VALUE "--------".
     05   FILLER                         PIC X(10)
          VALUE "-----".
     05   FILLER                         PIC X(11)
          VALUE "------".
     05   FILLER                         PIC X(11)
          VALUE "------".
     05   FILLER                         PIC X(10)
          VALUE "---------".

01   LARGEST-COMMISSION-LINE.
```

```
      05   FILLER                         PIC X(5)    VALUE SPACES.
      05   FILLER                         PIC X(26)
           VALUE "LARGEST SALES COMMISSION: ".
      05   O-LARGEST-COMMISSION           PIC 999.9   VALUE ZERO.
      05   FILLER                         PIC X(25)
           VALUE " SOLD BY SALESMAN NUMBER "
      05   LARGE-COMM-SALESMAN            PIC 9(8).   VALUE ZERO.

01  OUTPUT-RECORD.
      05   FILLER                         PIC X(5)    VALUE SPACES.
      05   O-SALESMAN-NUMBER              PIC 9(8).
      05   FILLER                         PIC X(5)    VALUE SPACES.
      05   O-SALES-QUOTA                  PIC ZZ9.99.
      05   FILLER                         PIC X(4)    VALUE SPACES.
      05   O-SALES                        PIC ZZ9.99.
      05   FILLER                         PIC X(5)    VALUE SPACES.
      05   O-BONUS-FACTOR                 PIC .99.
      05   FILLER                         PIC X(8)    VALUE SPACES.
      05   O-COMMISSION                   PIC ZZ9.99.

01  COMMISSION                           PIC 999V99  VALUE ZERO.
01  BLANK-LINE                           PIC X(132)  VALUE SPACES.

01  LARGEST-COMMISSION                   PIC 999V99  VALUE ZERO.

01  ADDITIONAL-SALES                     PIC 999V99 VALUE ZERO.

PROCEDURE DIVISION.

MAIN-LINE-ROUTINE.

    OPEN INPUT INPUT-SALES-FILE
        OUTPUT REPORT-FILE.
    WRITE REPORT-RECORD FROM HEADING-LINE-1
        AFTER ADVANCING TO-TOP-OF-PAGE.
    WRITE REPORT-RECORD FROM BLANK-LINE
        AFTER ADVANCING 2 LINES.
    WRITE REPORT-RECORD FROM BLANK-LINE
        AFTER ADVANCING 2 LINES.
    WRITE REPORT-RECORD FROM BLANK-LINE
        AFTER ADVANCING 1 LINES.
    WRITE REPORT-RECORD FROM HEADING-LINE-2
        AFTER ADVANCING 1 LINES.
    WRITE REPORT-RECORD FROM HEADING-LINE-3
        AFTER ADVANCING 1 LINES.
    WRITE REPORT RECORD FROM HEADING-LINE-4
        AFTER ADVANCING 1 LINES.
    WRITE REPORT-RECORD FROM BLANK-LINE
        AFTER ADVANCING 1 LINES.

    READ INPUT-SALES-FILE
        AT END PERFORM NEXT-LINE.
    MOVE SALESMAN-NO TO LARGE-COMM-SALESMAN.
    PERFORM PROCESS-RECORD
        UNTIL SALESMAN-NO EQUALS ZERO.
    MOVE LARGEST-COMMISSION TO O-LARGEST-COMMISSION.

    WRITE REPORT-RECORD FROM BLANK-LINE
        AFTER ADVANCING 2 LINES.
    WRITE REPORT-RECORD FROM LARGEST-COMMISSION-LINE
        AFTER ADVANCING 1 LINES.
    CLOSE INPUT-SALES-FILE
        REPORT-FILE.
    STOP RUN.
```

```
PROCESS-RECORD.

    MOVE SALESMAN-NO TO O-SALESMAN-NUMBER.
    READ INPUT-SALES-FILE
        AT END PERFORM NEXT-LINE.
    SUBTRACT SALES-QUOTA FROM SALES
        GIVING ADDITIONAL-SALES.
    IF ADDITIONAL-SALES IS NEGATIVE

        MOVE ZERO TO COMMISSION

    ELSE

        MULTIPLY ADDITIONAL-SALES BY BONUS-FACTOR
            GIVING COMMISSION
    IF COMMISSION IS GREATER THAN LARGEST-COMMISSION

        MOVE O-SALESMAN-NUMBER TO LARGE-COMM-SALESMAN
        MOVE COMMISSION TO LARGEST-COMMISSION

    MOVE SALES-QUOTA TO O-SALES-QUOTA.
    MOVE SALES TO O-SALES.
    MOVE BONUS-FACTOR TO O-BONUS-FACTOR.
    MOVE COMMISSION TO O-COMMISSION.

    WRITE REPORT-RECORD FROM OUTPUT-RECORD
        AFTER ADVANCING 1 LINES

    READ INPUT-SALES-FILE
        AT END PERFORM NEXT-LINE.

NEXT-LINE.
```

Inventory Control Processing

The more difficult programming problem that follows will also be solved with the top-down approach.

A manufacturing company with facilities for stocking up to 50 inventory items decides to develop a program that will maintain inventory control records on a small minicomputer. For each stock item, the following data is to be stored on the computer:

1. The stock number (a five-digit integer)
2. A count of the number of items on hand
3. The total year-to-date sales count
4. The item price
5. The date (month and day) of the last order (a four-digit integer of the form MMDD, which is zero if there is no outstanding order)
6. The number of items ordered (zero if there is no outstanding order)

Initially, the data for eight stock items is stored in computer memory. A transaction, which is an input request to the inventory program, consists of two records: a header record and a data record. The format of these is shown in Table 4.2 (d indicates a digit).

Table 4.2 *Input Formats for Inventory Control Program*

Header Record

Col 1	Col 7	Col 16
↓	↓	↓
ddddd	ddddd	ddddd
Transaction Number	Transaction ID	Stock Number

Data Record

Col 1	Col 8	Col 15	Col 20	Col 25	Col 30
↓	↓	↓	↓	↓	↓
ddd.dd	ddd.dd	dddd	dddd	dddd	dddd
New Price	Price	On-Hand	Date	Volume	Number Wanted

The transaction ID number indicates the type of transaction and the processing required, as shown in Table 4.3.

Table 4.3 *Description of Transactions*

Keywords	Meaning
PRICE (4)	Change stock item price
NEW (3)	Place new order
RECEIVED (2)	Process orders received
PURCHASE (1)	Process purchase orders
ADD (0)	Add new stock item

A set of typical transactions is shown in Table 4.4.

Transaction header and data records are input until all transactions are processed. After this occurs, the current inventory is output with the format shown in Table 4.5 (d indicates a digit).

Now that the specific input and output formats have been defined, we will describe how data is to be stored. Six tables, each of size 50, will be used to store the stock number, the number of items on hand, the total year-to-date sales count, the item price, the date of the last order, and the number of items ordered, respectively. The tables are shown in Fig. 4.15.

Table 4.4 *Typical Inventory Control Transactions*

Trans. No.	Header Record Trans. ID	Stock No.	Data Record	
			Price	*On-Hand*
4	ADD (0)	25812	16.25	59
9	ADD (0)	12348	40.90	4
			No. Wanted	
1	PURCHASE (1)	26541	5	
5	PURCHASE (1)	56912	11	
8	PURCHASE (1)	37999	1	
			Volume	
10	RECEIVED (2)	23451	56	
			Date	*Volume*
3	NEW (3)	12345	0310	20
15	NEW (3)	99976	0815	9
16	NEW (3)	61094	0901	23
20	NEW (3)	00123	0314	5
			New Price	
30	PRICE (4)	08234	15.95	
31	PRICE (4)	10234	20.00	
35	PRICE (4)	01236	19.25	
36	PRICE (4)	19456	45.10	

Table 4.5 *Inventory Control Report Formats*

Col 16
INVENTORY REPORT

Col 1 ↓	Col 8 ↓	Col 16 ↓	Col 25 ↓	Col 52 ↓	Col 40 ↓
Stock Number	*Items on Hand*	*Total Sales*	*Item Price*	*Date last Ordered*	*Quantity Ordered*
ddddd	ddddd	ddddddd	ddd .dd	dddd	dddd
.	.	.	.	.	.
.	.	.	.	.	.
.	.	.	.	.	.
ddddd	ddddd	ddddddd	ddd .dd	dddd	dddd

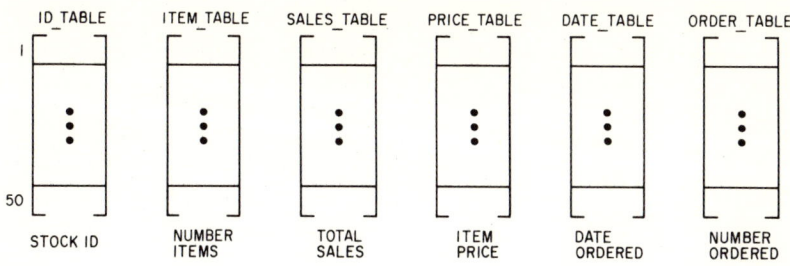

Fig. 4.15 *Inventory control program tables*

We are now in a position to apply the top-down approach by first defining the highest, most general problem description, as follows:

$$P_0$$
Maintain Inventory Control Records

The next level of refinement, showing some high-level logic, appears in Example 4.9.

Example 4.9 *P_1 Level of Inventory Control Program*

```
PROGRAM: TO PROCESS INVENTORY

    while more data do the following

        input record

        if legal transaction then

            process transaction

        else

            print 'invalid transaction'

    print inventory report

end * program *
```

We can now refine level P_1, showing more details and some of the specific error conditions that may occur, as shown in Example 4.10.

From level P_2, it is clear that the logic to search for the stock number, the logic to process each transaction type, and the inventory report can all be defined as subroutines or modules. If you recall, the structuring of a program into subroutines or modules involves the problem-solving technique known as "subgoaling." Using this approach, the overall pro-

Example 4.10 *P₂ Level of Inventory Control Program*

```
PROGRAM: TO PROCESS INVENTORY
    while more data do the following
        input record
        search for stock number
        if a match then
            --determine type transaction
            if add items then
                print 'duplicate stock number'
            else
                if legal transaction then
                    process transaction
                else
                    print 'invalid transaction'
        print inventory report
    end * program *
```

gram will be comprised of a main program module and 7 subroutines or modules.

It is convenient to express this overall program design in terms of a structure chart, or hierarchy chart, which is very similar in appearance to a common business organizational chart. A structure chart illustrates the relationships of the subroutines, or modules, in the system, each subfunction being described in terms of a verb or phrase, as shown in Fig. 4.16.

We will continue our top-down approach by further refining level P_2 to level P_3. Level P_3 will show the detailed logic that enables the program to be written directly into a programming language. The subroutine or

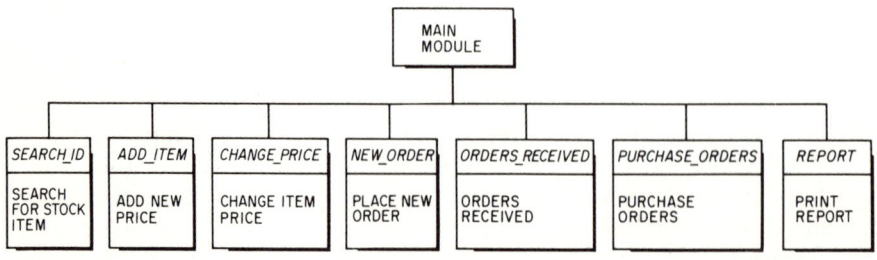

Fig. 4.16 *Structure chart of inventory control program*

module names that have been previously defined in the structure are referenced, along with the parameters passed from the main module to the called module. Level P_3 is shown in Example 4.11.

The logic of each subroutine called from the main program includes the following: search for item stock number, process price changes, new orders, orders received, purchase orders, adding orders, and outputting

Example 4.11   *P_3 Level of Inventory Control Program*

```
PROGRAM: TO CONTROL INVENTORY

Constant Definitions

  CURRENT_NUM_ITEMS = 8
  MAX_NUM_ITEMS = 50
  NUM_TRANSACTIONS = 100


  for TRANSACTION = 1 to NUM_TRANSACTIONS do the following

      input TRANS_NUM,TRANS_ID,STOCK_NUM
      call SEARCH_ID ( * giving *    STOCK_NUM,
                                     CURRENT_NUM_ITEMS,
                       * receiving * STOCK_ENTRY_FOUND)

      if STOCK_ENTRY_FOUND less or equal CURRENT_NUM_ITEMS then


          case TRANS_ID of

            0:
                print 'STOCK NUMBER',STOCK_NUM,'ALREADY EXISTS'

            4:

                call CHANGE_PRICE ( * giving * STOCK_ENTRY_FOUND)

            3:

                call NEW_ORDER ( * giving * STOCK_ENTRY_FOUND)

            2:

                call ORDERS_RECEIVED ( * giving * STOCK_ENTRY_FOUND)

            1:

                call PURCHASE_ORDERS ( * giving * STOCK_ENTRY_FOUND)

          end  * case *

        else

          if TRANS_ID = 0 then

              call ADD_ITEM ( *   giving *    STOCK_ENTRY_FOUND,
                                              STOCK_NUM,
                                              MAX_NUM_ITEMS,
```

```
                              *  updating  *  CURRENT_NUM_ITEMS)

             else

                print 'STOCK ITEM',STOCK_NUM,'NOT FOUND'

      call REPORT ( * giving * CURRENT_NUM_ITEMS)

      **  SEARCH_ID        SUBROUTINE (SEE EXAMPLE 4.12) **
      **  CHANGE_PRICE     SUBROUTINE (SEE EXAMPLE 4.13) **
      **  NEW_ORDER        SUBROUTINE (SEE EXAMPLE 4.14) **
      **  ORDERS_RECEIVED  SUBROUTINE (SEE EXAMPLE 4.15) **
      **  PURCHASE_ORDERS  SUBROUTINE (SEE EXAMPLE 4.16) **
      **  ADD_ITEM         SUBROUTINE (SEE EXAMPLE 4.17) **
      **  REPORT           SUBROUTINE (SEE EXAMPLE 4.18) **

    end  * program *
```

the inventory report on the printer. Each is described separately in Examples 4.12 through 4.18, respectively.

You will note that the main program inputs the transaction header record, while the appropriate transaction processing module inputs the associated data record.

The subroutine in Example 4.12 searches for the stock number in table ID_TABLE and passes back "STOCK_ENTRY_FOUND" as the stock number position in the table, if found. If the stock number is not found, STOCK_ENTRY_FOUND will be set to CURRENT_NUM_ITEMS (number of stock items) plus 1. Control is then returned to the main program module.

Example 4.12   *Subroutine to Search for Stock Number*

```
subroutine SEARCH_ID ( * using * STOCK_NUM,
                                  CURRENT_NUM_ITEMS,
                       * giving * STOCK_ENTRY_FOUND)

   set STOCK_ENTRY_FOUND  to  1

   while (STOCK_ENTRY_FOUND less or equal CURRENT_NUM_ITEMS) or
         (ID_TABLE(STOCK_ENTRY_FOUND) not equal STOCK_NUM) do following

      add 1  to  STOCK_ENTRY_FOUND
   return

end  * subroutine *
```

The subroutine in Example 4.13 inputs and replaces the new stock item price in PRICE_TABLE. Control is then returned to the main program module.

The subroutine in Example 4.14 inputs and updates the new order information for a stock item in tables DATE_TABLE and ORDER_TABLE. Control is then returned to the main program module.

Example 4.13 *Subroutine to Process Price Change*

```
subroutine CHANGE_PRICE ( * using * STOCK_ENTRY_FOUND)

    input NEW_PRICE
    set PRICE_TABLE(STOCK_ENTRY_FOUND)  to  NEW_PRICE
    return

end  * subroutine *
```

Example 4.14 *Subroutine to Process New Orders*

```
subroutine NEW_ORDER ( * using * STOCK_ENTRY_FOUND)

    input ORDER_DATE, VOLUME
    set DATE_TABLE(STOCK_ENTRY_FOUND)  to   ORDER_DATE
    set ORDER_TABLE(STOCK_ENTRY_FOUND) to   VOLUME
    return

end  * subroutine *
```

The subroutine in Example 4.15 inputs and processes orders received in tables DATE__TABLE, ORDER__TABLE, and ITEMS__ TABLE. Control is then returned to the main program module.

The subroutine in Example 4.16 inputs and processes purchase orders by updating information in tables ITEMS__TABLE and SALES__TABLE. Control is then returned to the main program module.

Example 4.15 *Subroutine to Process Orders Received*

```
subroutine ORDERS_RECEIVED (* using * STOCK_ENTRY_FOUND)

    input VOLUME_RECEIVED
    set DATE_TABLE(STOCK_ENTRY_FOUND) to 0
    set ORDER_TABLE(STOCK_ENTRY_FOUND) to 0
    add VOLUME_RECEIVED  to  ITEMS_TABLE(STOCK_ENTRY_FOUND)
    return

end  * subroutine *
```

Example 4.16 *Subroutine to Process Purchase Orders*

```
subroutine PURCHASE_ORDERS( * using * STOCK_ENTRY_FOUND)

  input NUM_ITEMS_WANTED
  if NUM_ITEMS_WANTED greater than ITEMS_TABLE(STOCK_ENTRY_FOUND) then
     print 'ORDER CAN NOT BE FILLED'

  else

     add NUM_ITEMS_WANTED  to  SALES_TABLE(STOCK_ENTRY_FOUND)
     subtract NUM_ITEMS_WANTED from ITEMS_TABLE(STOCK_ENTRY_FOUND)
  return

end  * subroutine *
```

The subroutine in Example 4.17 inputs and adds new stock item information to tables ID_TABLE, ITEMS_TABLE, SALES_TABLE, PRICE_TABLE, DATE_TABLE, and ORDER_TABLE. Control is then returned to the main program module.

The subroutine in Example 4.18 prints the inventory report with the formats shown in Table 4.5. Control is then returned to the main program module.

Example 4.17   *Subroutine to Process Add Items*

```
subroutine ADD_ITEM ( * giving *    STOCK_ENTRY_FOUND,
                                     STOCK_NUM,
                                     MAX_NUM_ITEMS,
                      * updating * CURRENT_NUM_ITEMS)

   input ITEM_PRICE, ITEMS_ON_HAND
   if CURRENT_NUM_ITEMS less than MAX_NUM_ITEMS then

      set ID_TABLE(STOCK_ENTRY_FOUND)      to   STOCK_NUM
      set ITEMS_TABLE(STOCK_ENTRY_FOUND)   to   ITEMS_ON_HAND

      set SALES_TABLE(STOCK_ENTRY_FOUND)   to   0
      set PRICE_TABLE(STCK_ENTRY_FOUNDS)   to   ITEM_PRICE

      set DATE_TABLE(STCK_ENTRY_FOUNDS)    to   0
      set ORDER_TABLE(STCK_ENTRY_FOUNDS)   to   0

      add 1 to CURRENT_NUM_ITEMS

   else

      print 'STOCK FACILITIES EXCEEDED'
   return

end  * subroutine *
```

Example 4.18   *Subroutine to Print Inventory Report*

```
subroutine REPORT( * using * CURRENT_NUM_ITEMS)

   print                    inventory report'
   print one blank line
   print 'STOCK   ITEMS   TOTAL    ITEM    DATE     QUANTITY'
   print 'NUMBER ON HAND SALES    PRICE   ORDERED ORDERED'
   print '......  ........  .....      .....   .......  ........'

   for CURRENT_ENTRY = 1 to CURRENT_NUM_ITEMS do the following

      print ID_TABLE(CURRENT_ENTRY),
            ITEM_TABLE(CURRENT_ENTRY),

            SALES_TABLE(CURRENT_ENTRY),
            PRICE_TABLE(CURRENT_ENTRY),

            DATE_TABLE(CURRENT_ENTRY),
            ORDER_TABLE(CURRENT_ENTRY)
   return

end  * subroutine *
```

In summary, the fundamental idea behind the top-down approach to program problem solving is to break large problems down into smaller individual problems. We first define the problem in terms of the highest general definition and then refine it in successive levels of detail until a final level of program design is achieved.

One of the most powerful characteristics of this approach is that the logic details for subroutines or modules are unnecessary until the main module is completely designed. Once this is accomplished, each module is designed as an independent problem.

5

Debugging

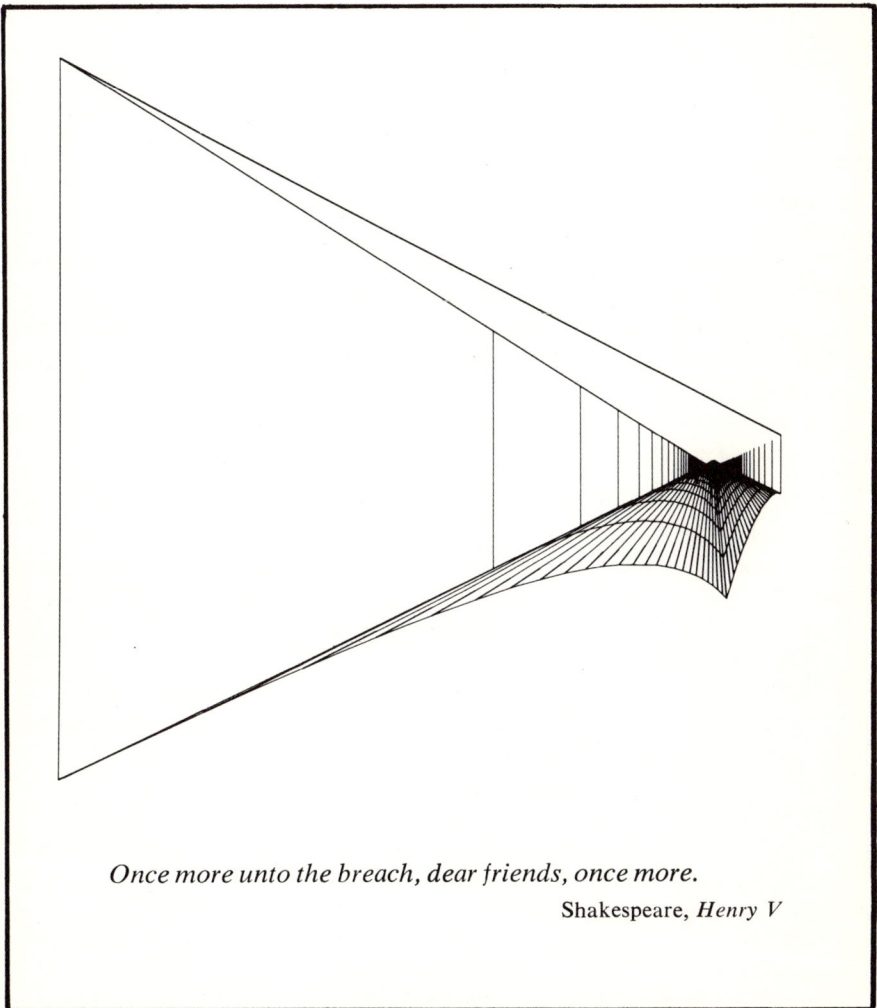

Once more unto the breach, dear friends, once more.

Shakespeare, *Henry V*

5.1   A Bug Is a Problem to Be Solved

All programs are written with the hope that they will work the first time they are executed. However, programmers soon learn that this happy event is the exception rather than the rule. It has been said that a program that runs correctly the first time is as rare as a hole-in-one. The beginning programmer comes to realize that the "debugging" stage[1] can be the most difficult part of the programming process since it puts his or her logical ability to its most extreme test.

What type of logical process is required during the debugging process? Is there any correlation between it and the basic processes involved in problem solving? Most people familiar with the two are inclined to believe that they are basically similar. A close analysis, however, will conclusively prove that they are identical!

Why is there so little literature on debugging? The answer lies in the lack of established and agreed-upon methods for the process. Perhaps it also reflects the fact that everyone expects a programmer to know how to debug his programs automatically. No doubt, this is partly true, since the programmer is the person most familiar with his own programs. Nevertheless, not all programmers are effective debuggers.

One explanation is that if a programmer is not disciplined in program solving, he will not be effective in problem definition, solution planning, *or* debugging. We will thus look upon debugging as a special problem-solving situation and apply the same primitive thinking tools to it that we did to other solutions.

Fallacy of the Last Bug

One common failing in program debugging, particularly true of inexperienced programmers, is to assume that a program is completely debugged before being absolutely certain. One cause of this failing is the "game effect" discussed in Chap. 3. The game effect induces initial interest in ferreting out bugs, but it can also interfere with the process when debugging starts to seem more like work than play.

The more errors we find in a program, the more we believe that it is now "working." As time passes, we may even start going around in circles, testing the familiar logic paths of the program even though they have already been tested and verified.

This mental block is caused by the relative ease of revisiting areas in the program that we have already fixed, for they represent *achievements!* In fact, this vicious circle may be so distracting that we stop looking for new bugs; that represents *hard work.*

---

[1] The origin of the word "bug" in programming dates back to the early development of computers when a moth caused a computer to malfunction. Removal of the insect solved the problem, and the term "debugging" evolved.

In conclusion, it is imperative to pursue the debugging process diligently, always assuming that some bug or other will eventually surface.

5.2   The Anatomy of a Program Bug

Symptoms, Operations, and the Bug

We have seen that general problem solving consists of givens, operations, and a goal and involves manipulating those givens to reach the goal in a series of operations. In programming, the givens, operations, and goal correspond to input, processing, and output. In debugging, however, the correspondence is to symptoms, operations, and the bug, as shown in Fig. 5.1.

Fig. 5.1   *Debugging as a problem-solving process*

Debugging is a special case of general problem solving with a specific goal in mind—to find the bug and eliminate it. The symptoms that help us locate the bug include diagnostic error messages, incorrect output, input/output errors, abnormal program termination, and, worst of all, the destruction of computer memory to the extent that it ceases to function. Operations in debugging consist of those primitive thinking tools that allow us to deduce the bug from the symptoms.

Since we have characterized program debugging as a problem-solving situation, we must consider the constraints that may impede our successful location of a bug, just as we do in general problem solving. Debugging constraints include such factors as a lack of memory printouts and logic traces, unfamiliarity with the program logic, and unavailability of the computer.

We should also be reminded of the psychological forces that intimately influence our success or failure in general problem solving, for they are also at work in debugging. These forces both aid and inhibit our ability to debug programs. Inhibitive ones include the game effect, ignoring known symptoms, a defensive ego, lack of imagination, failure to brainstorm, imagining unnecessary constraints, and failure to use incubation.

To find a program bug, we must first develop a plan of attack to isolate it. One simple approach is to simulate the logic flow of the program with paper and pencil. Regardless of our approach, however, debugging requires an analysis of the cause-effect relationships that lead us from the symptoms to the bug. As an example, consider the payroll programming problem of Chap. 2. It entailed the processing of payroll stubs for a group of employees, in which the inputs were the employee name, the number of hours worked for a particular week, and the hourly pay rate. Output consisted of printed employee pay checks and stubs.

Assume that the payroll program had been written in a slightly different way, as shown in Example 5.1. We now experience an error condition characterized by an unreasonably small check amount, apparently not solely due to inflation!

Example 5.1 *Program to Process Payroll Stubs (with Bug)*

```
PROGRAM: TO PROCESS PAYROLL

Constant Definitions

  MAX_SECURITY  = 1200
  RATE1_INCOME  = .07
  RATE2_INCOME  = .15
  SOCIAL_RATE   = .65
  CREDIT_RATE   = .05
  STOCK_RATE    = .10
  INCOME_BRACKET 100.0
  NUM_EMPLOYEES = 50

for CURRENT_EMPLOYEE = 1 to NUM_EMPLOYEES do the following

    input EMPLOYEE_ID, HOURS_WORKED, PAY_RATE, TOTAL_SECURITY
    set GROSS_PAY  to  PAY_RATE * HOURS_WORKED

    if GROSS_PAY less or equal to INCOME_BRACKET then
       set INCOME_TAX  to  RATE1_INCOME * GROSS_PAY
    else
       set INCOME_TAX  TO  RATE2_INCOME * GROSS_PAY

    if TOTAL_SECURITY less than MAX_SECURITY then
       set SOCIAL_SECURITY  to  SOCIAL_RATE * GROSS_PAY
       add SOCIAL_SECURITY  to  TOTAL_SECURITY

       if TOTAL_SECURITY greater than MAX_SECURITY then
          subtract (TOTAL_SECURITY - MAX_SECURITY)
                  from SOCIAL_SECURITY

       set TOTAL_SECURITY  to  MAX_SECURITY

    else

       set SOCIAL_SECURITY  to 0

    set CREDIT_UNION  to  CREDIT_RATE * GROSS_PAY
    set STOCK_PLAN  to  STOCK_RATE * GROSS_PAY

    set OTHER_DEDUC  to  INCOME_TAX + SOCIAL_SECURITY + CREDIT_UNION +
                         STOCK_PLAN

    set NET_PAY  to  GROSS_PAY - OTHER_DEDUC

    print '   RATE      HOURS      GROSS      NET'
    print one blank line
```

```
print PAY_RATE, HOURS_WORKED, GROSS_PAY, NET_PAY
print one blank line

print ' TAX     FICA      OTHER     EMPLOYEE ID'
print INCOME_TAX, SOCIAL_SECURITY, OTHER_DEDUC, EMPLOYEE_ID

end * program *
```

If we start examining each logical step from the beginning of a program, our knowledge of the state relationships continually increases, as shown in Fig. 5.2. For example, state 2 informs us that the input data is valid, and state 3 lets us know that the gross pay calculation is correct. State 5, however, reveals that the social security tax calculation is excessive. Upon examining the actual program, we discover that the social security deduction is 65 percent, not the specified 6.5 percent, a clerical error apparently introduced while designing the program.

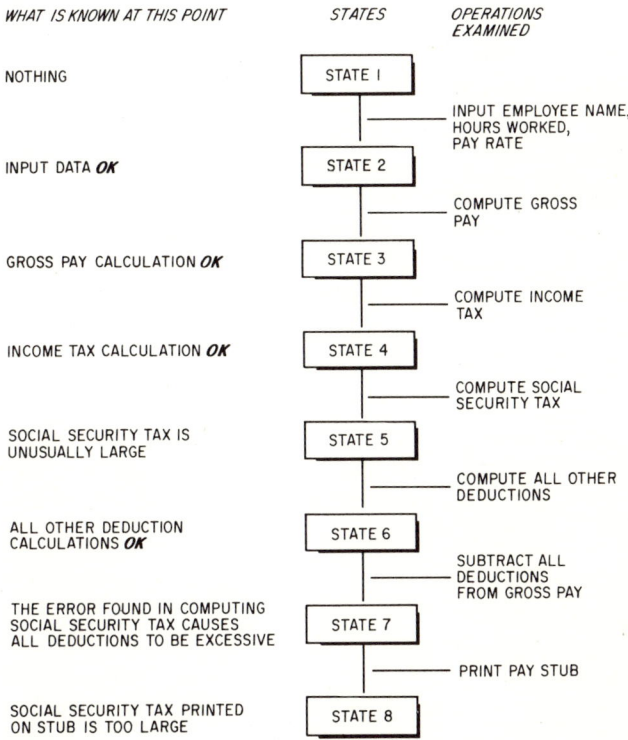

Fig. 5.2 *States of debugging flow in payroll program*

Even after finding this error, it is wise to verify the rest of the program. Only by state 8 can we be sure that the social security calculation is the only error in the program that caused the short pay check. Moving the decimal point two places to the left in the social security calculation eliminates the bug.

The direction of our search in this example was "forwards" in the sense that we worked from the beginning of the program until the error was found. We will also use other primitive thinking tools to discover errors in a program, including working in reverse—from the symptom to the cause.

5.3   Sources of Errors

Garbage In—Garbage Out (GIGO)

The popular belief that the computer affords a panacea for all ills overlooks the crucial importance of input. The programming phrase "garbage in–garbage out"—commonly contracted to GIGO—refers to the fact that if input is not reliable and accurate, the output will not be reliable or accurate either—thus "garbage." A wise defensive measure is to build limit checks into a program. For example, in the payroll program, checks might be limited to a maximum of $2000.00.

Clerical Errors

Clerical errors commonly occur during the transcription of a program or data when coding or preparing input. Examples are the faulty transposition of adjacent digits, such as writing 4978 instead of 4798, confusing the letter O and the number 0, accidentally leaving out program statements, and placing logical operations out of order.

Faulty Logic

Faulty logic refers to the use of the wrong program solution method, placing logical program statements out of order, missing logic, and irrelevant logic. These errors stem primarily from inadequate problem definition and solution planning.

Poor Assumptions

Poor assumptions range from oversimplification of the original problem to overcomplication of it. In both cases, the original problem cannot be solved. The remedy once again is careful adherence to adequate problem definition.

Computational Errors

The representation of numbers in a computer often causes problems in terms of precision, overflow, and truncation. For example, the total number of significant digits associated with a given number is called its *precision;* when an attempt is made to exceed this number, an error occurs. Overflow errors occur whenever an attempt is made to compute a number outside the allowable numerical range of a variable.

Errors also occur when two numbers of unequal precision are numerically combined in some manner. The right-hand digits of the smaller-resolution number will be truncated, producing unexpected results.

Other computational errors include incorrect expressions and the misuse of operand precedence. Operand precedence refers to the order of mathematical operations, typically, exponentiation, multiplication, division, addition, and subtraction. Suppose we have coded the following programming expression

$$X = A/B \times C + D$$

for $A = 4$, $B = 4$, $C = 2$, and $D = 5$. If we assume the above operand precedence, the result will be as follows:

$$X = 4/(8) + 5 = 5.5$$

If the operand precedence is division before multiplication, however, the result will change to the following:

$$X = (4/4) \times 2 + 5 = 7$$

Inconsistent Units

Inconsistent units refer to the use of dissimilar data units. Output data units may differ from input units with no logical programming conversion to resolve the inconsistency. For example, input units may be in quarts and output units in gallons.

Boundary Conditions

Boundary conditions refer to the existence of certain sensitive boundaries whose violation can result in an error. One example is the use of invalid operations, leading to an invalid problem solution or even catastrophic error.

As an example of a nonprogramming boundary condition, consider the following mathematical reasoning:

(1) Since $2 + 2 = 4$, let $A = 2$ and $B = 2$

(2) $A = B$ (Given)

(3) $A^2 = AB$ (Multiply both sides by A)

(4) $A^2 - B^2 = AB - B^2$ (Subtract B^2 from both sides)

(5) $(A + B)(A - B) = B(A - B)$ (Factor both sides)

(6) $\dfrac{(A + B)(A - B)}{(A - B)} = \dfrac{B(A - B)}{(A - B)}$ (Divide both sides by $A - B$)

(7) Hence, $A + B = B$ or $2 + 2 = 2$ (Cancel $A - B$ on both sides)

The reason for the absurd result is that we performed an illegal arithmetic operation. In step (6) we divided both sides of the equation by zero, an operation undefined in mathematics. The same error will occur whenever a program statement attempts to divide a number or variable by zero or execute any other invalid instruction.

Other examples of boundary violations include the following:

1. Invalid arguments in system functions
2. Overlaying program statements with input data
3. Outputting program statements
4. Attempting to assign a variable a value outside its range
5. Insufficient main memory
6. Multiple use of a single variable
7. Addressing memory beyond the physical limits of the main memory
8. Addressing locations outside a specific program or data area
9. Timing problems
10. Deleting too much information from a stack, resulting in stack underflow
11. Calling a nonexistent module
12. Starting off with the wrong index in a loop
13. Going beyond the legal bounds of a loop index

In conclusion, boundary conditions are a frequent source of programming bugs. To avoid such errors, we must be very sensitive to program bounds and make sure that they are never violated—an alertness called "defensive programming."

5.4   Prescription for Debugging

The following prescriptions are organized similarly to previous ones but relate exclusively to debugging. The general idea is presented first and is often followed by a computer example.

PRESCRIPTION 1
Get Your Bearings

It is essential to understand the characteristics of bugs and all associated symptoms before attempting to track them down. You should attempt to isolate each culprit by asking questions such as "Is the bug actually in my program or outside of it?," "Is it a hardware bug?," "Is it a timing bug?" Once you have a handle on the general problem area, you will be in a better position to establish a specific course of action.

Bugs are frequently due to something *outside* the program, for example, using the wrong compiler option. However, be warned that almost every bug is *in your program,* not in the compiler or hardware.

As you gain more insight about symptoms, it is advisable to re-evaluate your position in the debugging process from time to time. It is all too easy to lose sight of what you have learned and where you are going.

As an example of this evaluation process, consider the output from the sales commission problem of Chap. 4 that is shown in Table 5.1.

Table 5.1 *Output from Sales Commission Program*

SALES COMMISSION REPORT

Salesman Name	Sales Quota	Actual Sales	Bonus Factor	Commission Paid
34512472	$400	$380	.20	$ .00
12147983	$300	$340	.18	$ 7.20
59346821	$250	$375	.25	$31.25
.	.	.	.	.
.	.	.	.	.
.	.	.	.	.

Largest Sales Commission: $31.75 sold by salesman 59346821

It is apparent from the data that the salesman with the largest commission has been determined correctly, but the commission paid him—$31.25—does not agree with the amount printed as the largest sales commission, $31.75 (the tens digit differs). The first obvious question is, "Is this a hardware or programming error?" The printer may have output the wrong character because of an electrical or mechanical failure, which could be either a temporary or permanent condition. On the other hand, the printer might be working correctly, and a bug *does* exist in the program. "Getting your Bearings" refers to the fact that if you are to get a handle on the problem, you must first *relax* and think about the symptoms. By asking such questions and narrowing down the general trouble area, you will be in a better position to develop a plan of attack and examine specific causes.

PRESCRIPTION 2

Determine If Bug Is Consistent

Determining whether a bug is consistent is very important; it can eliminate many nonprogramming errors over which you have no control. A bug may not be repeatable for several reasons:

1. Operator error (for example, mounting wrong tape)
2. Timing errors (for example, Real Time Systems)

3. Transient errors in hardware
4. Invalid data

It might be advisable to re-execute the program for the problem in Prescription 1 to see if the symptom repeats itself; the hardware might have momentarily "burped" because of a power fluctuation. In this event, it would be a waste of time to search for a programming error. When errors are determined not to have human causes, the following procedure is advised:

1. Rerun the program
2. Run hardware diagnostics
3. Attempt to find the bug in the program
4. If all else fails, notify the vendor (bug may be in software or hardware)

PRESCRIPTION 3
Courage Is Grace Under Pressure

This prescription states that we must momentarily set aside our ego in the debugging process, for debugging requires the utmost of our logical abilities. When finding a bug proves difficult, we often start blaming the hardware, even though it is seldom the culprit. It is human nature to unburden ourselves of our frustrations and project them momentarily on something else. What could be more convenient than the hardware? Common "whipping boys" are the compiler, the assembler, the operating system, or the fact that the program is someone else's.

It is essential to concentrate solely on the problem at hand, disdaining facile excuses. It is sometimes helpful to make a game of the process and pretend that you are a detective in search of a suspect.

PRESCRIPTION 4
Zero In on Simple Clues First

This prescription states that it is most productive to consider the most obvious candidates for bugs first and zero in on these. Should you disregard this prescription, you will find yourself going down blind alleys; eventually you will have to consider the obvious symptoms anyway.

When you have a set of equally suspected candidates, a corollary to this prescription states that you should examine the suspects in terms of their increasing level of difficulty. It is more economical to examine and eliminate simple suspects than to tackle the more difficult ones first.

PRESCRIPTION 5

Solve Mystery with Contradiction and Elimination

The method of contradiction is fundamental to debugging. To understand it, we must first understand the associated terminology.

A contradiction occurs when two statements, taken together, cannot both be true or false. As an example, consider the two statements: "Today is Tuesday" and "Today is not Tuesday." There is no way that both of these statements can be true, nor can they both be false; obviously, any given day must either be Tuesday or not Tuesday.

Beware of confusing a contradiction with contrary statements such as "Today is Tuesday" and "Today is Wednesday." These statements are not contradictory because on Friday, for instance, both statements are false.

The important characteristic of contradictory statements is that one of them must be true and the other false. Or, to put it another way, there is *one and only one* true statement.

One of the basic logical methods for proving something is to make two statements: (1) the statement that you wish to prove, and (2) the contradiction of the statement that you wish to prove. Thus provided with two statements, one of which is false and the other true, you proceed to prove that the contradictory statement is false. If you are able to do so, you then know that the remaining statement must be true and that you have proven the thing you started out to prove.

Contradiction provides a simple path for analysis and significantly reduces the number of possible bugs. Usually referred to as the "process of elimination," it is a very powerful primitive thinking tool for problems involving many variables or mountains of data.

Non-Computer Example

Consider the following murder mystery, in which we are provided with the thirteen clues listed in Table 5.2 to help us identify the murderer. The suspect can definitely be proven to be the murderer if we carefully analyze these clues.

When the number of things that we must consider to solve a problem becomes too great, it is often helpful to draw up a chart organizing the information, relationships, and constraints in a compact form. Such a chart will also help us visualize the problem and keep track of each new conclusion reached.

In Table 5.2, six statements refer to rooms, with one reference to an adjacent room. Since it is important to visualize this information, let us construct the diagram of rooms shown in Fig. 5.3. Once we have this diagram, we list the various characteristics associated with each room, as shown in Fig. 5.4. Mr. Jones is assigned room 700 and a Ford automobile.

Table 5.2   *Clues to the Identity of the Murderer*

1. Hopewell lives in 704.
2. Smith always wears tan.
3. The man in the black suit has red hair.
4. Holman has gray hair.
5. Jones lives in 700 and drives a Ford.
6. The man in 706 is bald.
7. The man who lives next to the Cadillac driver wears blue.
8. The man in 708 has brown hair.
9. The man in the gray suit drives a Rambler.
10. The man in 704 drives a Chevy.
11. The man in the blue suit has black hair, and his initials are J. H.
12. The suspect wears a brown suit and doesn't drive a Lincoln.
13. V. Higgins doesn't drive a Rambler and doesn't live in 702.

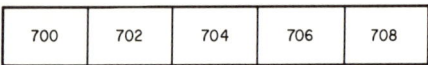

Fig. 5.3   *Room data*

ROOM	700	702	704	706	708
MAN	JONES		HOPEWELL		
HAIR				BALD	BROWN
CAR	FORD		CHEVY		
SUIT					

Fig. 5.4   *Given problem relationships*

Hopewell is assigned Room 704 and a Chevy. Moreover, the man in room 706 is bald, and the man in room 708 has brown hair.

Now note that statement 12 tells us that the suspect wears a brown suit. If we can determine who wears this suit, we can identify the suspect.

At this point, we begin to form statement pairs, that is, statements that relate two pieces of information, for example:

1. Smith and tan suit
2. Black suit and red hair
3. Gray suit and Rambler
4. Blue suit, black hair, and initials J. H.

We will now use these statements to obtain new information. Clearly, Smith and the tan suit can only be assigned to rooms 702, 706, or 708. From statement 13, however, we know that Higgins must be in room 706 or 708 because he can't be in room 702. Since Holman has gray hair,

ROOM	700	702	704	706	708
MAN	JONES	HOLMAN	HOPEWELL		
HAIR	RED	GRAY	BLACK	BALD	BROWN
CAR	FORD		CHEVY		
SUIT	BLACK		BLUE		

Fig. 5.5 *Additional problem relationships established*

he can't be the person in rooms 706 or 708 because one of these is bald and the other has brown hair. Holman is therefore in room 702. Smith and Higgins must be in rooms 706 and 708, although not necessarily in that order. Likewise, the initials J. H. cannot belong to either Holman or Higgins, because Holman has gray hair and Higgins, who is in room 706 or 708, must either be bald or have brown hair. The initials J. H. therefore belong to Hopewell, who wears a blue suit and has black hair. Jones, who lives in room 700, must be the one who has red hair and wears a black suit. We fill in this additional information as shown in Fig. 5.5.

Since Higgins or Smith must live in rooms 706 or 708, let us arbitrarily assume that Higgins is in 706 and Smith in room 708. From clue 7 (the man who lives next to the Cadillac driver wears blue) and the fact that Hopewell is in room 704 and wears a blue suit, we know that the Cadillac must be associated with rooms 702 or 706. Let us assume that the Cadillac is associated with room 702. Since Higgins in 706 does not own a Rambler (clue 13), he must therefore own the Lincoln. Thus the Rambler must belong to Smith in room 708. But since the Rambler is associated with a gray suit, this supposition contradicts previous information that Smith wears a tan suit. The Cadillac must therefore belong to Higgins of room 706 and the Rambler to Smith of room 708. However, the fact that a gray suit is associated with the Rambler contradicts the information that Smith wears a tan suit. The Rambler must therefore belong to Holman of room 702, who wears a gray suit. Since we know Smith of room 708 wears a tan suit, Higgins must therefore wear the brown suit and is thus our murder suspect. The completed diagram, which solves the problem, is shown in Fig. 5.6. We have successfully applied the primitive thinking tool of contradiction and elimination to locate the murderer from many clues.

ROOM	700	702	704	706	708
MAN	JONES	HOLMAN	HOPEWELL	HIGGINS	SMITH
HAIR	RED	GRAY	BLACK	BALD	BROWN
CAR	FORD	RAMBLER	CHEVY	CADILLAC	LINCOLN
SUIT	BLACK	GRAY	BLUE	BROWN	TAN

Fig. 5.6 *Solution of problem*

Computer Example

Let us now consider a programming problem that illustrates the use of contradiction and elimination. The ACE Electric Company charges its customers according to the following rate schedule:

8 cents a kilowatt-hour for electricity used up to the first 300 kwh
6 cents a kilowatt-hour for electricity used up to 600 kwh
5 cents a kilowatt-hour for electricity used up to 1000 kwh
3 cents a kilowatt-hour for electricity used over 1000 kwh

The output from the program consists of a three-column table listing the customer number, kilowatt hours used, and the charge. It also includes the number of customers, total hours used, and the total charges. The program itself is shown in Example 5.2. The input data is shown in Table 5.3. The output data is shown in Table 5.4.

Table 5.3   *Input Data to Energy Program*

Customer Number	Kilowatt-Hours Used
125	725
390	600
231	115
492	48
921	1200
954	395

Table 5.4   *Output Results from Energy Program*

Customer Number	Hours Used	Charge
125	725	48.25
390	600	42.00
231	115	9.20
492	48	3.84
921	1200	66.00
954	395	29.70
TOTAL CUSTOMERS	TOTAL HOURS	TOTAL CHARGES
6	3083	198.99

Paper and pencil calculations, however, show that the total charges should be $200.99 rather than the $198.99 that the computer output. Since there is clearly a bug in the program, let us approach the problem as a "mystery" and apply contradiction and elimination to find it.

Example 5.2 *Kilowatt Program (with Bug)*

```
PROGRAM: TO COMPUTE UTILITY BILL

Constant Definitions

   KWH300  = 300
   KWH600  = 600
   KWH400  = 400
   KWH1000 = 1000
   RATE1   = 0.08
   RATE2   = 0.06
   RATE3   = 0.05
   RATE4   = 0.03
   TOTAL_CUSTOMERS = 100

   set TOTAL_HOURS to 0
   set TOTAL_CHARGES to 0
   print 'CUSTOMER NUMBER  HOURS USED  CHARGE'
   print '...............  ..........  ......'

   for CURRENT_CUSTOMER = 1 to TOTAL_CUSTOMERS do the following
       input CUSTOMER_NUMBER, NUM_HOURS
       set SAVED_HOURS to NUM_HOURS
       add NUM_HOURS to TOTAL_HOURS

       if NUM_HOURS greater than KWH300 then

           set CHARGE to RATE1 * KWH300
           subtract KWH300 from NUM_HOURS

           if NUM_HOURS greater than KWH300 then

               add RATE2 * KWH300 to CHARGE
               subtract KWH300 from NUM_HOURS

               if NUM_HOURS greater than KWH400 then

                   add RATE3 * KWH300 to CHARGE
                   subtract KWH300 from NUM_HOURS
                   add RATE4 * NUM_HOURS to CHARGE

               else

                   add RATE3 * NUM_HOURS to CHARGE

           else

               add RATE2 * NUM_HOURS to CHARGE

       else

           set CHARGE to RATE1 * NUM_HOURS

       print CUSTOMER_NUMBER,SAVED_HOURS,CHARGE
       add CHARGE to TOTAL_CHARGES

   print 'TOTAL CUSTOMERS  TOTAL HOURS   TOTAL CHARGES'
   print  TOTAL_CUSTOMERS,TOTAL_HOURS,TOTAL_CHARGES

end * program *
```

The first check is to look at the other totals to see if they can give us a clue. We quickly determine that the total number of customers and total hours used were computed correctly and can therefore eliminate totalling as a source of error.

Another important check is to verify that the data has been input and printed correctly, for if not, the total charges will be incorrect (see the discussion on Garbage In–Garbage Out). Inspection reveals that the input data has been read and printed correctly, thus eliminating this possibility.

The only other possibilities are that the "totals" cumulation was incorrect *or* that the individual customer charges were not processed properly. An examination of the program logic that handles the cumulation of total charges shows that this calculation is merely a simple cumulation of individual customer charges previously calculated and that it has been performed properly.

Therefore, the only possible source of the bug is the individual customer charges. For the first input of 725 hours, the program correctly calculates the charge to be .08 cents times the first 300 hours, plus .06 cents times the next 300 hours, plus .05 cents times the remaining 125 hours, for a total charge of $48.24. The second input is also correct, for it is .08 cents times the first 300 hours, plus .06 cents times the remaining 300 hours, for a total charge of $42.00.

Continuing in this manner, we find no discrepancy until we get to the fifth input. The charges for hours up to 300 and 600 are processed correctly. However, for hours over 1000, .05 cents is multiplied by 300; it should be 400 to account for the incremental jump from 600 to 1000. Also, the number of hours subtracted at this point to lump hours over 1000 is 300; it should be 400. These two errors account for the $2.00 undercharge. The corrected program is shown in Example 5.3.

We have thus identified the bug by applying the method of contradiction and elimination, examining the symptoms and eliminating all conditions that might have caused the problem. To lighten the chore, we thought of ourselves as engaged in a "mystery" that required us to track the suspect down.

Example 5.3 *Kilowatt Program (Debugged)*

```
PROGRAM: TO COMPUTE UTILITY BILL

Constant Definitions

  KWH300   = 300
  KWH600   = 600
  KWH400   = 400
  KWH1000  = 1000
  RATE1    = 0.08
  RATE2    = 0.06
  RATE3    = 0.05
  RATE4    = 0.03
  TOTAL_CUSTOMERS = 100

  set TOTAL_HOURS to 0
  set TOTAL_CHARGES to 0
  print 'CUSTOMER NUMBER   HOURS USED   CHARGE'
  print '................  ..........   ......'
```

```
for CURRENT_CUSTOMER = 1 to TOTAL_CUSTOMERS do the following
    input CUSTOMER_NUMBER, NUM_HOURS
    set SAVED_HOURS to NUM_HOURS
    add NUM_HOURS to TOTAL_HOURS

    if NUM_HOURS greater than KWH300 then

        set CHARGE to RATE1 * KWH300
        subtract KWH300 from NUM_HOURS

        if NUM_HOURS greater than KWH300 then

            add RATE2 * KWH300 to CHARGE
            subtract KWH300 from NUM_HOURS

            if NUM_HOURS greater than KWH400 then

                add RATE3 * KWH400 to CHARGE
                subtract KWH400 from NUM_HOURS
                add RATE4 * NUM_HOURS to CHARGE

            else

                add RATE3 * NUM_HOURS to CHARGE

        else

            add RATE2 * NUM_HOURS to CHARGE

    else

        set CHARGE to RATE1 * NUM_HOURS

    print CUSTOMER_NUMBER,SAVED_HOURS,CHARGE
    add CHARGE to TOTAL_CHARGES
print 'TOTAL CUSTOMERS  TOTAL HOURS   TOTAL CHARGES'
print  TOTAL_CUSTOMERS,TOTAL_HOURS,TOTAL_CHARGES

end * program *
```

PRESCRIPTION 6

Don't Ignore Known Symptoms

One of the most common difficulties in debugging is the failure to take advantage of known symptoms. Symptoms can manifest themselves as any of the following:

1. Invalid output
2. Endless logic loops
3. Program variables with unexpected values
4. Clobbered program statements
5. Clobbered data
6. Unexpected logic paths

It is absolutely essential for you to think of symptoms as "clues" that tell you something about the bug; they should always be treated as friends. Your efforts and time will be wasted if you don't do so.

PRESCRIPTION 7
Incubate When Stuck on Bug

In Chap. 4, we demonstrated the importance of "incubating"— putting a problem aside—whenever you experience a mental block in problem definition and planning. Incubating is also important in debugging a program, for we often reach an impasse in which all the logical reasoning in the world will not help.

Finding a bug is often a binary process, that is, it is either found or not found. In problem definition or planning, there is usually a wider choice of alternatives to foster the creative process. In debugging, your frustrations in reaching the goal have a single source—the bug. Incubation becomes all the more important for resolving your difficulties.

It is vital at such times to forget about the bug completely. Do something relaxing, like sleeping, having a drink, playing tennis, watching television. Sometimes incubation can occur on the way to the coffee machine; it often takes no more than a momentary distraction to make the bug pop out.

PRESCRIPTION 8
Apply Binary Search to Capture Culprit

One of the most unrewarding and time-consuming methods for identifying a program bug is the "random search," in which you haphazardly probe various unrelated areas in a program with no thought of a consistent plan of attack. With a great deal of luck, this method might prove fruitful; far more often your efforts will be wasted.

A better approach is to employ an organized binary search, as illustrated in Fig. 5.7.

A binary search is initiated by partially executing the program and examining the variables and data after reaching its approximate midpoint. Execution is halted by placing a "stop" instruction in the program or by other methods provided by the program language, the hardware, or the software support. If no error is found, the bug must be located somewhere after this midpoint.

The next step is to examine the second portion of the program logic between points B and G in Fig. 5.7. The program is re-executed and examined at the approximate midpoint of this region, or point F. If an eror is encountered, we know that the bug is located between points B and F rather than points F and G. We must therefore re-execute the program and

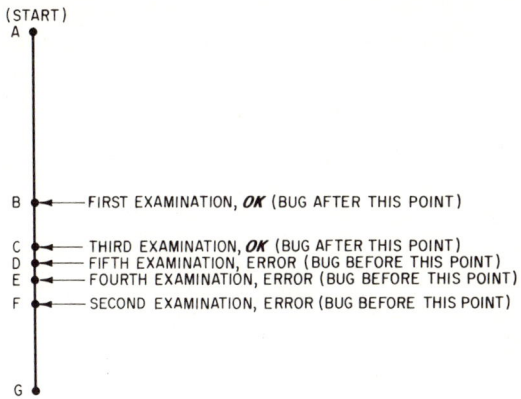

PROGRAM LOGIC SEQUENCE

(START)

A

B — FIRST EXAMINATION, *OK* (BUG AFTER THIS POINT)

C — THIRD EXAMINATION, *OK* (BUG AFTER THIS POINT)
D — FIFTH EXAMINATION, ERROR (BUG BEFORE THIS POINT)
E — FOURTH EXAMINATION, ERROR (BUG BEFORE THIS POINT)
F — SECOND EXAMINATION, ERROR (BUG BEFORE THIS POINT)

G

Fig. 5.7 *Binary search for the bug*

examine point C, which lies between points B and F. If no error is found, we know that the bug is located between points C and F rather than points B and C.

In the fourth try, we re-execute the program and examine the mid-point between points C and F, or point E. If there is an error in this region, as there is in Fig. 5.7, we know that the bug must reside between points C and E. A fifth try examines the program between points C and E, or point D. If an error is found, as one is in Fig. 5.6, we know that the bug must reside between points C and D. Continuing the binary search in this manner, we are able to narrow down the possible problem area very quickly.

Binary searching is a special case of finding bugs by using contradiction and elimination, but it also employs working forwards and backwards, combined.

PRESCRIPTION 9

Blessed Is He Who Takes Nothing for Granted

This prescription says that you should not assume that any particular portion of a program is working unless you can conclusively prove that fact. Otherwise your efforts are doomed to frustration.

For example, consider finding a bug by using a nonbinary search, as illustrated in Fig. 5.8. Let us assume that you believe that the region between points F and G does not contain the bug. Two reasons for your belief could be your realization that this area has "worked" in the past and your suspicion that the bug is very close to the beginning of the program.

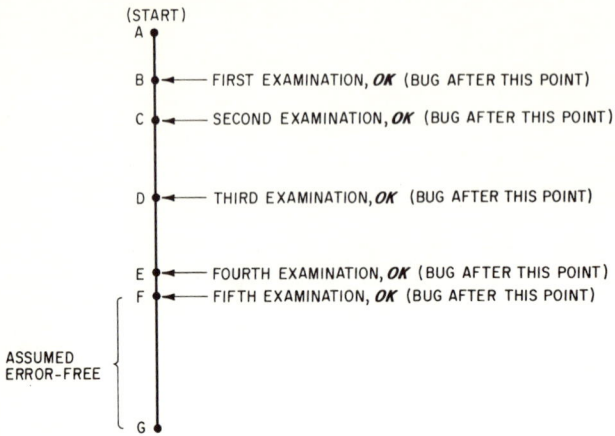

Fig. 5.8 *Search with predefined assumptions*

You start your search in a forwards direction from point A, examining the program at point B. Finding no bug there, you continue in a forwards direction, moving closer and closer to the region between points F and G, which was initially assumed to be error-free. You then stop the program at point C, find no bug there, and continue the process still without finding the bug.

By the fifth examination, you realize that your original assumption was incorrect and that the bug must reside between points F and G. You have thus wasted five trials because of that assumption. A better approach would have been not to assume *anything*. A binary search would have located the bug after a few trials.

PRESCRIPTION 10

Test from the Top

This debugging prescription has its counterpart in programming, as was discussed in Sec. 4.2 of Chap. 4 that emphasized the importance of approaching a problem from the "top."

One way to understand top-down testing is to discuss another form of testing, known as "bottom-up" testing, first. In many ways the exact opposite of top-down testing, bottom-up testing is illustrated in Fig. 5.9. Assume that the program or system being tested is fairly complex, involving a set of "modules" or subroutines.

In bottom-up testing, individual modules at the lowest level of the logical tree structure—A3, A4, A5, A6, B2, and C3—are tested in a

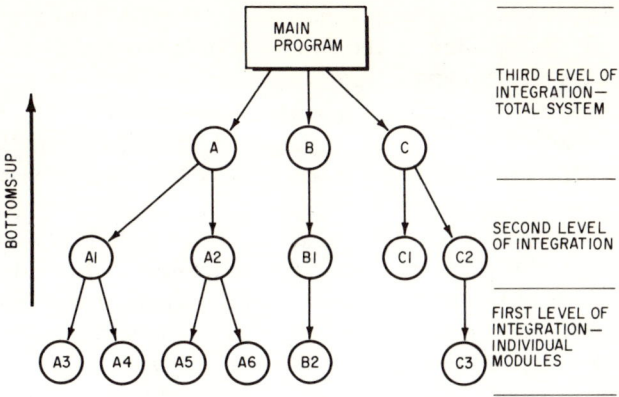

Fig. 5.9 *Bottoms-up testing approach*

stand-alone fashion. Once verified to be bug-free, these components are integrated into the next higher level of the logical structure, that is, modules A1, A2, B1, C1, and C2. When this level has also been verified to be bug-free, it is integrated into the next higher level, that is, modules A, B, and C. In the final stage of integration, all subsections are incorporated in the main program or module until the entire system has been tested.

The testing approach is thus from the "bottom" in the sense that we start with the lowest logical units and progressively integrate them into larger subsections of logic. The disadvantage of this approach is that it may take a long time to discover that our architectural design does not fit into the overall structure of the system. A more viable way to test a large program or system is to employ the "top-down" approach, as illustrated in Fig. 5.10.

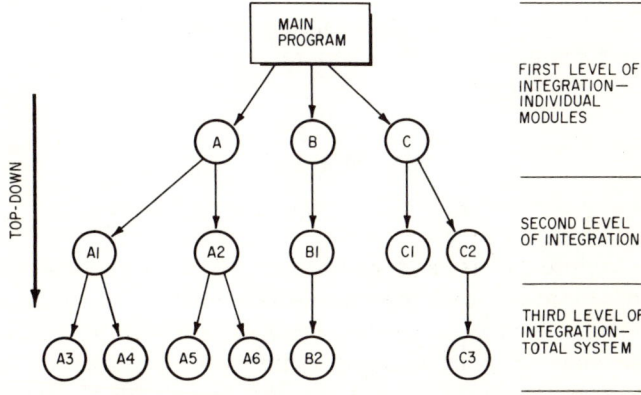

Fig. 5.10 *Top-down testing approach*

Modules or subroutines associated with the top level of the program design—A, B, and C—are tested first and then integrated into the main program. Once the top-level logical interfaces have been verified to be correct, we may integrate the next lower level—modules A1, A2, B1, C1, and C2—into the previous level—modules A, B, and C. Finally, the lowest modules are integrated into the second level of integration, resulting in a completely tested system.

Top-down testing has several advantages over bottom-up testing. First of all, it can be accomplished in parallel with top-down design. Once the highest level of the program has been designed, it can be coded and tested, giving us a running "prototype" model of the system. Consequently, architectural bugs in the highest level can be found early on in program development.

An effective technique in top-down testing is called "stubbing." In stubbing, lower-level modules are initially created with just enough logic to permit them to be interface-tested with higher-level modules. These stub modules are replaced with the real logic as soon as this is developed. For example, the function of a lower-level module may be to find the average of N numbers, passed from a higher-level module. The temporary stub module may simply pass back a specific predefined number to the higher-level module or print a message.

Another advantage of top-down testing is that it usually makes system bugs easier to identify. Since the lower modules are "supportive" to the higher modules, and since the higher modules are examined first, most major design bugs will be found more quickly.

Yet another advantage of this approach is that it distributes testing time more evenly throughout the overall development of the program and does not delay it until the very end when some sensitive areas of the design may have been forgotten.

A final advantage to top-down testing is that it implies a subgoaling approach to problem solving. As you recall, subgoaling is a very powerful tool for simplifying the problem-solving process. Breaking a problem down into parts and treating each separately makes the overall debugging process more manageable.

PRESCRIPTION 11

Use Inference to Narrow Suspects

Since inference is a logical method of generating additional information—in this case, symptoms—from information already given, it can be of great service in debugging. As an illustration, consider the following problem. You are asked to write a program that will input "N" data items and calculate the average and the standard deviation. The standard deviation provides additional insight into the characteristics of the data, for it represents a measure of the dispersion around the average and a measure of the data variation. The mathematical formulas for the average and standard deviation are:

$$\text{Average} = \frac{1}{N} \sum_{i=1}^{N} \text{Data i}$$

$$\text{Standard deviation} = \sqrt{\frac{1}{N} \sum_{i=1}^{N} (\text{Data i} - \text{Average})^2}$$

    The program consists of a main module to input the data and print and store it in a table called DATA__TABLE. Input is terminated after N data items have been input. The main module calls a subroutine, STATISTICS, which computes the average and the standard deviation. Examples 5.4 and 5.5 show the main module and the subroutine logic, respectively.

Example 5.4 *Statistical Problem (Main Module)*

```
PROGRAM: TO COMPUTE AVERAGE AND STANDARD DEVIATION

Constant Definitions

   N = 5

   print 'SCORES:'
   for CURRENT_NUMBER = 1 to N do the following

      input DATA_TABLE(CURRENT_NUMBER)
      print DATA_TABLE(CURRENT_NUMBER)

   call STATISTICS( * giving *      N,
                    * receiving *   AVERAGE,
                                    STND_DEVIATION)

   print 'AVERAGE:',AVERAGE,'STANDARD DEVIATION:',STND_DEVIATION

end   * program *
```

Example 5.5 *Statistical Problem (Subroutine)*

```
subroutine STATISTICS ( * using *    N,
                        * giving *   AVERAGE,
                                     STND_DEVIATION)

   set AVERAGE   to   0
   set STND_DEVIATION   to   0
   set CURRENT_DATA_ITEM to 1

   while  CURRENT_DATA_ITEM less or equal N do the following
      add (DATA_TABLE(CURRENT_DATA_ITEM) / N)  to  AVERAGE

   divide  AVERAGE by N
```

```
for CURRENT_DATA_ITEM = 1 to N do the following
                                                    2
    add (DATA_TABLE(CURRENT_DATA_ITEM)  - AVERAGE)     to
        STND_DEVIATION

    set STND_DEVIATION  to square root of (STND_DEVIATION / N)
    return

end  * subroutine *
```

Input to this program consists of the five data items shown in Table 5.5.

Table 5.5 *Input Data*

Item	Data
1	94
2	57
3	69
4	82
5	100

When the program is first executed, a strange thing occurs. It does not print out the average and the standard deviation but apparently enters an "endless loop." As a matter of fact, the computer operator must intervene to terminate the program.

One might infer that the bug resides in the logic involving looping. An inspection of the program determines that the only suspect places are in the main routine when the data is input or in the subroutine STATISTICS when the average and the standard deviation are calculated.

One cause for the endless loop might be a failure to input the five data items; without them, the program would indeed loop forever. Since the complete input data is printed before the loop is entered, however, we can infer that the data has been input correctly and that the bug is not located in the main module. It is also evident, by inspection, that the parameter interfaces between the main module and the subroutine STATISTICS are correct. We can thus infer that the bug must reside somewhere in the subroutine.

An examination of the first loop in the subroutine STATISTICS, which calculates the average, reveals that the CURRENT__DATA __ITEM associated with DATA__TABLE is not incremented between repeated cumulations of data items. Since the loop will never terminate without this incrementation, its absence is the cause of the problem. We have thus found the bug, which can be fixed by adding the logic needed to increment the CURRENT__DATA__ITEM, using a FOR loop.

When the program is now re-executed, the average and the standard deviation prove to be 16.08 and 66.2337, respectively. Since the printed average, 16.08, is absurdly small relative to the input data, we realize that we must have another bug in the program.

When we take the time to calculate the true average and standard deviation with paper and pencil, we find them to be 80.4 and 15.8, respectively, a more reasonable result. We thus infer that there must be a bug in the averaging calculation of the subroutine STATISTICS. Upon examining the logic of this subroutine, we notice that the expression

add (DATA_TABLE (CURRENT_DATA_ITEM)/N) to AVERAGE

divides each data item by the number of data items. This calculation is suspicious, for we merely want to cumulate the data items and divide the result by the total number of data items. When we eliminate this faulty division, as shown in Example 5.6, and re-execute the program, the program now works, giving us the exact values for the average and the standard deviation that we obtained with pencil and paper. Our bugs have thus been tracked down with inference, a very powerful primitive thinking tool indeed!

Example 5.6 *Statistical Program (Subroutine)*

```
subroutine STATISTICS ( * using *   N,
                        * giving *   AVERAGE,
                                     STND_DEVIATION)

   set AVERAGE  to  0
   set STND_DEVIATION  to  0

   for CURRENT_DATA_ITEM = 1 to N do the following
      add (DATA_TABLE(CURRENT_DATA_ITEM))  to  AVERAGE

   divide  AVERAGE by N

   for CURRENT_DATA_ITEM = 1 to N do the following
                                                      2
      add (DATA_TABLE(CURRENT_DATA_ITEM)  - AVERAGE)    to
          STND_DEVIATION

   set STND_DEVIATION  to square root of (STND_DEVIATION / N)
   return

end  * subroutine *
```

PRESCRIPTION 12

Misery Enjoys Company—Brainstorm!

In Chap. 3 we studied a very powerful approach to problem solving—brainstorming—in which we share our flow of ideas with other individuals.

Brainstorming can also be a great help in debugging. If our imagination fails, nothing could be wiser than to discuss our problem with others, especially colleagues working on the same project who have a general understanding of the goals of the program.

Even if you are the sole programmer, it may be a good idea to seek advice from others. Free of your mistakes, they may have a positive contribution to make. Often, merely describing the characteristics of the bug to them will help you gain a new perspective on the problem. It might be just what the doctor ordered to get you on the right track.

PRESCRIPTION 13

Dump Program and Trace Logic Sparingly

Storage Dumps

Another aid to the debugging process is the simple storage dump, in which the programmer dumps the contents of memory at execution time—typically to the printer. A storage dump may be useful for printing the contents of memory in areas where a bug is suspected. Be warned, however, that dumps may produce only garbage for you—a big waste of paper.

Dumps are analogous to police dragnets in that they may catch all suspects or, on the other hand, none. Since dumps may be expensive in terms of computer time, your time, and paper, you must take care to assure that they will provide additional insight into program errors.

It is very important for dumps to be formatted in a readable form; unstructured dumps only waste time. Table 5.6 illustrates a formatted dump in hexadecimal notation.

Table 5.6 *Storage Dump*

Dump address 100–127

Memory Address								
0100	02F6	0491	FFFF	4104	0000	49F2	0034	0123
0110	0954	034E	E285	034C	0F48	F4CE	0001	F218
0120	F623	EEF1	9821	0F56				

Logic Traces

Another useful debugging tool is the "trace." Traces print the contents of selected memory locations at specified times or whenever a specific logic sequence is encountered. Typically, a logic trace prints the contents of specified variables to the printer so that variable changes can be monitored in the program.

Table 5.7 illustrates the typical output of a trace program provided in a high-level programming language.

Table 5.7 *Outputs from Program Trace*

Program Statement	Trace variables X, Y Current value of variables	
0100	X=2.78	Y=3.414
0110	X=4.98	Y=3.414
0340	X=25.2	Y=.003

PRESCRIPTION 14

Fish for the Bug with Hooks

When a computer manufacturer does not provide a trace facility, you may create your own by systematically inserting additional logic in your program. This technique is known as placing "hooks" in your program. Once the program is debugged, the hooks are removed.

As an example of this useful technique, consider the following problem. It assumes that your computer language does not provide the square root function, and you must write your own. You are asked to write a subroutine that will be passed a number from the main program for which you are to determine the square root and pass the result back to the calling sequence, as follows:

Main
Program   Subroutine:

call SQROOT(N,X) --------→ SUBROUTINE SQROOT(N,X)
                                        .
                                        .
                                        .
                                   RETURN

where N = the number for which the square root is to be computed and X = the result passed back to the main program.

A simple method for finding the square root of a number is called Newton's Method. Assume X to be an estimate, arrived at arbitrarily, of the square root of the number N. If X is larger than the square root of N, then N/X is less than the square root of N, and vice versa. Since the multiplication of X and N/X is equal to N, the two numbers can serve as reciprocal estimates to determine the square root of N.

Newton's Method repeatedly replaces the estimate of X with the average of the reciprocal estimates, as follows

$$X = .5(X + N/X)$$

until the absolute difference of the reciprocal estimates (X − N/X) is found.

A program based on this method is shown in Example 5.7. The initial guess for X is assumed to be 1 in all cases. Also, the accuracy for the square root must be within four decimal places, or .0001.

Example 5.7   *Finding Square Roots (First Attempt)*

```
subroutine: SQROOT( * using *  N,
                    * giving * X)

   set  X to  1
   while absolute value of (X - N / X)
         not equal .0001 do the following

      set  X to .5 * X + N / X

      * debug hook *
      print X
   return

end  * subroutine *
```

Notice that after the repeated value of X is calculated, a "hook" is placed in the program to print the current value of X, thus allowing us to trace the value of this variable as an aid in debugging.

When we execute the program for the first time, the value of N for which we want the square root is 171. The program output is shown in Table 5.8.

Our hook is telling us that the square root of N is not converging and that there therefore must be a bug in the program. When we examine the program, we discover that we are not really averaging the reciprocal estimates, for the current value is being effectively expressed as .5X + N/X when it should be .5(X + N/X). The program is corrected accordingly (see Example 5.8).

When we now re-execute the program, we obtain the output shown in Table 5.9.

Looking up the square root of 171 in mathematical tables, we discover to our delight that we have found the correct one, or 13.152946. The program fails to terminate, however, and has to be stopped by the computer operator. Obviously, it is not abiding by the stopping rule, which we

Table 5.8 Outputs from First Run of Square Roots Program

173.5	18.601075
87.747118	18.601075
45.845134	18.601075
26.696141	18.601075
19.828408	18.601075
18.639050	18.601075
18.601114	18.601075
18.601075	18.601075
18.601075	18.601075
18.601075	18.601075
18.601075	18.601075
18.601075	18.601075
18.601075	18.601075
18.601075	18.601075
18.601075	18.601075
18.601075	

Table 5.9 Outputs from Second Run of Square Roots Program

87	13.152946
44.494253	13.152946
24.191198	13.152946
15.671280	13.152946
13.355291	13.152946
13.154479	13.152946
13.152947	13.152946
13.152946	13.152946
13.152946	13.152946
13.152946	13.152946
13.152946	13.152946
13.152946	13.152946

Example 5.8 Finding Square Roots (Second Attempt)

```
subroutine: SQROOT( * using *  N,
                    * giving * X)

   set X to  1
   while absolute value of (X - N / X)
         not equal .0001 do the following

      set X to .5 * (X + N / X )

      * debug hook *
      print X
   return

end  * subroutine *
```

must now re-examine. Upon doing so, we notice that the decision check for program termination tests for a reciprocal difference that is *exactly* equal to the degree of approximation required, or .0001. It is clear that the convergence may never achieve this exact difference, and we must therefore change the logic to test for a difference that is less than or equal to the desired accuracy. A program reflecting this correction is shown in Example 5.9.

Example 5.9   *Finding Square Roots (Third Attempt)*

```
subroutine: SQROOT (* using *  N,
                    * giving * X)

    set X to   1
    while absolute value of (X - N  / X)
          greater than .0001 do the following

        set X to  .5 * (X + N / X)

        * debug hook *
        print X
    return

end  * subroutine *
```

The output from this version of the program is shown in Table 5.10.

Table 5.10   *Outputs from Third Run of Square Roots Program*

44.494253
24.191198
15.671280
13.355291
13.154479
13.152946

The subroutine is now completely debugged, for the required square root of N has been found and the program terminates properly. With time, however, it is discovered that on two occasions the program fails; the output symptoms on these occasions are shown in Table 5.11.

It is obvious from the symptoms that the program fails to check whether input values of N are larger than zero. The final program, in which this check in included and the debug hook removed, is shown in Example 5.10. It also incorporates an error message that is to be printed when invalid input is detected.

The beauty of using program "hooks" to aid us in debugging lies in the fact that we can put as many hooks in the program as we wish and easily remove them as soon as the program has been debugged.

Table 5.11 *Output from Program*

Case	Program Input Value of N	Output Results
#1	0	Message 'Data Overflow'
#2	−1	.5
		.25
		.125
		6.25E-2
		3.125E-2
		1.5625E-2
		7.8125E-3
		3.90625E-3
		1.953125E-3
		9.765625E-4
		4.882813E-4
		2.441406E-4
		1.220703E-4
		6.103516E-5

Example 5.10 *Finding Square Roots (Final Version)*

```
subroutine: SQROOT( * using *  N,
                    * giving *  X)

   set  X to  1
   if N greater or equal 0 then

      while absolute value of (X − N / X)
            greater than .0001 do the following

         set  X to  .5 * (X + N / X)
   else

      print N, 'IS LESS THAN ZERO'
   return

end  * subroutine *
```

PRESCRIPTION 15

Simulate with Paper and Pencil

A very powerful and inexpensive debugging technique, applicable even before a program is executed for the first time, is to desk-check the program logic with paper and pencil. You simulate the logic of your program by pretending that *you* are the computer and by keeping track of the input, output, and variable changes just as the logic would.

We used this very methodology when we "worked backwards" in Chap. 2. If you recall, in the water pitcher problem we tested our logical operations by keeping a record of the current contents of the small, large, and result pitchers on a piece of paper. In a desk check, the contents of the pitchers in the programming context become the contents of the "variables."

Consider the following program problem. You are asked to write a subroutine to compute the factorial N! of a single, arbitrary positive integer N. (N! = N × (N − 1) × (N − 2) × . . . × 2 × 1). As a special case, assume that 0! = 1.

The calling sequence from the main program to the subroutine FACTORIAL is designed to pass the value N as the first argument and then be returned the RESULT, N! This subroutine is shown in Example 5.11.

Example 5.11 *Finding Factorials (First Attempt)*

```
subroutine FACTORIAL (* using  *  N,
                      * giving *  RESULT)

   set RESULT  to   0
   set CURRENT_ITERATION to 1

   while CURRENT_ITERATION less or equal N  do the following

      set RESULT  to  CURRENT_ITERATION * RESULT
      add 1  to  CURRENT_ITERATION
   return

end  * subroutine *
```

We apply the method of desk-check simulation by first writing down on a piece of paper, in column form, the variables that will be modified in the program. The program is then simulated by writing down the changes to each variable at each step of the program logic. For example, suppose that the value of N that is passed to the subroutine is 7, and we therefore wish to find 7! The simulation of the logic is shown in Table 5.12.

Table 5.12 *Simulation of Program Logic*

N	RESULT	CURRENT_ITERATION
7	0	1
7	0	2
7	0	3
7	0	4
7	0	5
7	0	6
7	0	7

Since $7! = 7 \times 6 \times 5 \times 4 \times 3 \times 2 \times 1 = 5040$, we know that there must be a bug in the subroutine. Our paper and pencil simulation shows that the program will provide 0 as the result, and our efforts have paid off without wasting computer time.

When we examine the subroutine, we notice that we accidentally initialized the variable RESULT to zero; consequently, the final result will always be zero, as Table 5.12 shows. The variable result should instead be initialized to 1. When we correct this minor bug and simulate the program again, we obtain Table 5.13. The final subroutine logic is shown in Example 5.12.

In summary, we can use paper-and-pencil simulation to find many bugs. The technique is not only cheap, it can save a great deal of time.

Table 5.13 *Simulation of Program Logic*

N	RESULT	CURRENT_ITERATION
7	1	1
7	$1 = 1 \times 1$	2
7	$2 = 2 \times 1$	3
7	$6 = 3 \times 2$	4
7	$24 = 4 \times 6$	5
7	$120 = 5 \times 24$	6
7	$720 = 6 \times 120$	7
7	$5040 = 7 \times 720$	8

Example 5.12 *Finding Factorials (Final Version)*

```
subroutine FACTORIAL (* using  *  N,
                      * giving *  RESULT)

   set RESULT to 1
   set CURRENT_ITERATION to 1

   while CURRENT_ITERATION less or equal N  do the following

      set RESULT  to  CURRENT_ITERATION * RESULT
      add 1  to  CURRENT_ITERATION
   return

end  * subroutine *
```

PRESCRIPTION 16

Work Backwards from the Symptom

In Chap. 2 we saw that working backwards can be a very effective technique in the problem definition and solution plan stages. Working backwards, you will remember, focuses on the goal and determines the preceding logical operations that are required to reach it.

The same logical reasoning is applicable to debugging except that we start with the symptom and work backwards to the steps that produce it. This is a very powerful approach, for we do not have to start from the beginning of the program and trace through the whole logical structure. We merely backtrack from the symptom.

Consider the golf problem shown in Fig. 5.11. A golfer hits a ball with an initial velocity V (in feet per second) at an angle Z (in degrees) to the ground. The height H (in feet) of the ball above the ground is given in terms of the distance D (in feet) from the tee by the equation:

$$H = \frac{-16.1\ D^2}{V^2\ COS^2\ (Z)} + D\ TAN\ (Z)$$

where COS^2 = trigonometric cosine squared, and TAN = trigonometric tangent.

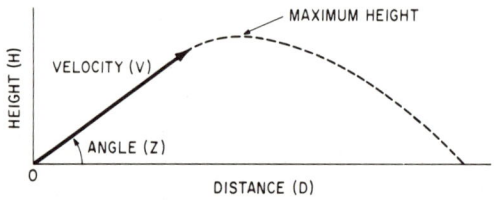

Fig. 5.11 *Golf problem*

You are asked to write a program that will input the values of the velocity, V, and the degrees, Z, and then compute and print the height, H, with one-foot increments in distance, D, until the ball hits the ground. The program should also calculate and print the maximum height the ball ever achieves.

The program you write to solve this problem is illustrated in Example 5.13, the inputs being as follows: V = 40 ft/s and Z = 15 deg.

When the program is executed for the first time, the following is printed:

Distance	Height
1	−.873429

Max. Height: 0

Since these figures indicate that the golf ball immediately hits the ground, there appears to be an error in the program. Even at an angle of only 15 degrees and a velocity of 40 ft/s, one would anticipate that the ball would at least get off the ground!

Example 5.13 *Golf Program (First Attempt)*

```
PROGRAM: TO COMPUTE GOLF STATISTICS

    set BALL_DISTANCE  to  1
    set BALL_HEIGHT  to  1
    set MAX_HEIGHT  to  0

    print 'DISTANCE      HEIGHT'
    input BALL_VELOCITY, BALL_DEGREES

    while BALL_HEIGHT greater or equal 0 do the following
                                                    2
        set BALL_HEIGHT  to  (-16.1 * (BALL_DISTANCE)   /
                        2                          2
            ((BALL_VELOCITY)  *   COS((BALL_DEGREES)  )) +
             BALL_DISTANCE   *  TAN(BALL_DEGREES)

        print BALL_DISTANCE, BALL_HEIGHT

        if BALL_HEIGHT less than MAX_HEIGHT then

            set MAX_HEIGHT  to  BALL_HEIGHT

        add 1 to BALL_DISTANCE

    print 'MAX.HEIGHT:', MAX-BALL_HEIGHT

end  * program *
```

An obvious suspect would be that part of the program which determines the height of the ball as a function of the distance, D. We will thus start working backwards from this point to locate the bug. An examination of the statement to calculate the height shows that it is coded correctly. We therefore backtrack from this point in the program to the steps that lead up to it.

Among these steps is the input of the variables in the equation, the velocity and the angle, which are previously initialized. We examine the input data and verify that the velocity is assigned as 40 and the angle as 15. However, intuition tells us that we should examine the angle more carefully. We dig into the programming manual supplied by the computer vendor and review the description of the system functions, Cosine and Tangent. To our surprise, these functions require arguments to be expressed in terms of radians, not degrees!

Aha! From our elementary trigonomety course we recall that radians are not the same as degrees, and it is clear that a conversion process must be performed to convert the degrees into radians. With a little research, we find that the number of radians in 180 deg is π, or approximately 3.14159. Simple algebra shows us that we can convert degrees to radians by multiplying the number of degress by the ratio of π to 180, as follows:

$$\text{Radians} = \text{Degrees} (\pi/180)$$

We can therefore eliminate the bug by inserting this conversion calculation into the program just after we input the number of degrees. The added program statement, where PI = 3.14159, is as follows:

multiply (PI/180) times BALL__DEGREES

When we make this correction and re-execute the program, we obtain the output shown in Table 5.14. The table shows that for a velocity of 40 ft/s and an angle of 15 deg, the distance of the ball from the tee when it hits the ground is 24 ft.

Table 5.14   *Output from Second Run of Golf Problem*

Distance	Height	Distance	Height
1	.257164	15	1.592520
2	.492758	16	1.526235
3	.706782	17	1.438281
4	.899237	18	1.328756
5	1.070121	19	1.97662
6	1.219435	20	1.044998
7	1.347180	21	.870763
8	1.453355	22	.674959
9	1.537959	23	.457585
10	1.500994	24	.218642
11	1.642469	25	$-4.187219E\text{-}2$
12	1.662354		
13	1.660679	Max. height: 0	
14	1.637434		

The program seems to be working now, for we have calculated the height of the ball as a function of the distance from home plate until the ball hits the ground. However, since the maximum height reached by the ball has been output as 0, there must be another program bug.

Again we work backwards to locate this bug. The only place where the maximum height is modified, other than when it is initially set to 0, is in the comparison between the present ball height and each one-foot increment of distance, D. When we locate this logic, we note that it consists of the following:

if BALL__HEIGHT less than MAX__HEIGHT then
set MAX__HEIGHT to BALL__HEIGHT

It is obvious that we have coded the if-conditional expression incorrectly, for the ball height will never be less than the maximum height, 0.

The value assigned to the maximum height will thus never change but will always be printed 0, as observed.

To correct this bug, the if-conditional expression must be rewritten as follows:

if BALL—HEIGHT greater than MAX—HEIGHT then
set MAX—HEIGHT to BALL—HEIGHT

When we change the program accordingly and re-execute it, the correct maximum height calculated and printed is 1.97662 ft, as can be verified in Table 5.14.

To assure ourselves that the program is working properly, we shall re-execute it with a different angle, 45 deg, but the same velocity, 40 ft/s. The output achieved is shown in Table 5.15 and indicates a completely debugged program. The final version of the program to solve the golf problem is shown in Example 5.14.

Table 5.15 *Output from Third Run of Golf Problem*

Distance	Height	Distance	Height
1	.979874	27	12.328859
2	1.919497	28	12.221984
3	2.818871	29	12.074859
4	3.677995	30	11.887484
5	4.496869	31	11.549860
6	5.275493	32	11.391985
7	6.013867	33	11.083850
8	6.711991	34	10.735486
9	7.369865	35	10.346861
10	7.987489	36	9.917987
11	8.564864	37	9.448862
12	9.101988	38	8.939488
13	9.598862	39	8.389864
14	10.055487	40	7.799990
15	10.471851	41	7.169865
16	10.847986	42	6.499491
17	11.183860	43	5.788867
18	11.479485	44	5.037993
19	11.734859	45	4.246869
20	11.949984	46	3.415495
21	12.124859	47	2.543872
22	12.259484	48	1.631998
23	12.353859	49	.679874
24	12.407984	50	− .312500
25	12.421859		
26	12.395484	Max. height: 12.421859	

Example 5.14 *Golf Program (Final Version)*

```
PROGRAM: TO COMPUTE GOLF STATISTICS

    set BALL_DISTANCE  to  1
    set BALL_HEIGHT  to  1
    set MAX_HEIGHT  to  0
    set PI to 3.14159

    print 'DISTANCE      HEIGHT'
    input BALL_VELOCITY, BALL_DEGREES
    multiply (PI / 180) times BALL_DEGREES

    while BALL_HEIGHT greater or equal 0 do the following
                                                      2
        set BALL_HEIGHT  to  (-16.1 * (BALL_DISTANCE)   /
                          2                    2
          ((BALL_VELOCITY)  *  COS((BALL_DEGREES)  )) +
            BALL_DISTANCE    *  TAN(BALL_DEGREES)

        print BALL_DISTANCE, BALL_HEIGHT

        if BALL_HEIGHT greater than MAX_HEIGHT then

            set MAX_HEIGHT  to  BALL_HEIGHT

        add 1 to BALL_DISTANCE

    print 'MAX.HEIGHT:', MAX-BALL_HEIGHT

end  * program *
```

In summary, working backwards is a very powerful primitive thinking tool for debugging. Its strength comes from the fact that we backtrack from the symptom where the bug resides.

Bibliography

Aron, Joel D. *The Program Development Process,* Addison-Wesley Publishing Company, 1974.

Barker, Stephen F. *The Elements of Logic,* McGraw-Hill Book Company, 1965.

Bartley, William Warren III. *Lewis Carroll's Symbolic Logic,* Clarkson N. Potter, Inc., Publishers, N.Y., 1977.

Beardsley, Monroe C. *Thinking Straight,* Prentice-Hall, Inc., Englewood Cliffs, N.J., 1975.

Bohl, Marilyn. *A Guide for Programmers,* Prentice-Hall, Inc., Englewood Cliffs, N.J., 1978.

Brody, Baruch A. *Logic,* Prentice-Hall, Inc. Englewood Cliffs, N.J., 1973.

Brooks, Frederick P. *The Mythical Man-Month,* Addison-Wesley Publishing Company, 1975.

Calter, Paul. *Problem Solving with Computers,* McGraw-Hill Book Co., 1973.

Coopersmith, Stanley. *Frontiers of Psychological Research,* W. H. Freeman and Co., 1966.

Davis, Gary A. *Psychology and Problem Solving,* Basic Books, Inc., Publishers, N.Y., 1973.

Davis, Gordon B. and Hoffman, Thomas R. FORTRAN, McGraw-Hill Book Company, 1978.

Dorf, Richard C. *Computer and Man,* Boyd & Fraser Publishing Co., San Francisco, Calif., 1974.

Friedman, Frank L. and Koffman, Elliot B. *Problem Solving and Structured Programming in FORTRAN,* Addison-Wesley Publishing Co., Inc., 1977.

Gardner, Martin. *AHA!,* Scientific American, Inc./W. H. Freeman and Company, 1978.

Hall, Calvin S. *Psychology,* Howard Allen, Inc., Cleveland, Ohio, 1960.

Haskell, Richard E. *FORTRAN Programming,* Scientific Research Associates, Inc., 1978.

Kieburtz, Richard B. *Structured Programming and Problem-Solving* with PL/1, Prentice-Hall, Inc., Englewood Cliffs, N.J., 1977.

Kieburtz, Richard B. *Structured Programming and Problem Solving with ALGO W,* Prentice-Hall, Inc., Englewood Cliffs, N.J., 1975.

Ledgard, Henry F. *Programming Proverbs,* Hayden Book Company, Rochelle Park, N.J., 1975.

Ledgard, Henry F. *Programming Proverbs for FORTRAN Programmers,* Hayden Book Company, Rochelle Park, N.J., 1975.

Ledgard, Henry F. and Chmura, Louis J. *FORTRAN with Style,* Hayden Book Company, Rochelle Park, N.J., 1978.

Ledgard, Henry F., Hueras, John F., and Nagin, Paul A. *PASCAL with Style,* Hayden Book Company, Rochelle Park, N.J., 1979.

Marateck, Samuel L. BASIC, Academic Press, 1975.

Neumann, John Von. *The Computer and the Brain,* Yale University Press, 1958.

Pirsig, Robert M. *Zen and the Art of Motorcycle Maintenance,* Bantam Books, 1974.

Rubinstein, Moshe F. *Patterns of Problem Solving,* Prentice-Hall, Inc., Englewood Cliffs, N.J., 1975.

Salmon, Wesley C. *Logic,* 2nd Ed., Prentice-Hall, Inc., Englewood Cliffs, N.J., 1973.

Wickelgren, Wayne A. *How to Solve Problems,* W. H. Freeman and Company, San Francisco, Calif., 1974.

Yourdon, Edward. *Techniques of Program Structure and Design,* Prentice-Hall, Inc., Englewood Cliffs, N.J., 1975.

Index

Index